# ASTRONOMY FOR YOUNG FOLKS

*ISABEL MARTIN LEWIS*

tion et mise en page : Culturea (Hérault, 34)
rouvez notre catalogue sur http://culturea.fr/
ntact : infos@culturea.fr
primé en Allemagne par Books on Demand Gmbh
sign typographique et layout : Derek Murphy et Reedsy
3N : 9791041985708
:e de parution : février 2024
livre :
a été composé avec une police open source libre de droits dessinée sur Open Fondry https://open-foundry.com/

# CHAPTER I. THE CONSTELLATIONS.

"Canst thou bind the sweet influences of the PleiadesOr loose the bands of Orion?Canst thou bring forth Mazzaroth in his seasonOr canst thou guide Arcturus with his sons?"

—BOOK OF JOB.

Who would not like to know the stars and constellations by their names and in their seasons as we know the birds and the trees and the flowers, to recognize at their return, year by year, Sirius and Spica, Arcturus and Antares, Vega and Altair, to know when Ursa Major swings high overhead and Orion sinks to rest beneath the western horizon, when Leo comes into view in the east or the Northern Crown lies overhead?

Often we deprive ourselves of the pleasure of making friends with the stars and shut our eyes to the glories of the heavens above because we do not realize how simple a matter it is to become acquainted with the various groups of stars as they cross our meridian, one by one, day after day and month after month in the same orderly sequence. When the robin returns once more to nest in the same orchard in the spring time, Leo and Virgo may be seen rising above the eastern horizon in the early evening hours. When the first snow flies in the late fall and the birds have all gone southward the belt of Orion appears in the east and Cygnus dips low in the west. When we once come to know brilliant blue-white Vega, ruddy Arcturus, golden Capella and sparkling Sirius we watch for them to return each in its proper season and greet them as old friends.

In the following pages we give for each month the constellations or star-groups that are nearest to our meridian, that is, that lie either due north or due south or exactly overhead in the early part of the month and the early part of the evening.

We do not need to start our study of the constellations in January. We may start at any month in the year and we will find the constellations given for that month on or near the meridian at the time indicated.

In using the charts or diagrams of the constellations, we should hold them in an inverted position with the top of the page toward the north or else remember that the left-hand side of the page is toward the east and the right-hand side of the page toward the west, which is the opposite of the arrangement for charts and maps of the earth's surface.

We should also bear in mind that the constellations are all continually shifting westward for the stars and the moon and the planets as well as the sun risedaily in the east and set in the west. This is due to the fact that the earth is turning in the opposite direction on its axis, that is from west to east. In twenty-four hours the earth turns completely around with respect to the heavens or through an angle of 360°, so in one hour it turns through an angle of 360° ÷ 24 or 15°. As a result the stars appear to shift westward 15° every hour. This is a distance about equal in length to the handle of the Big Dipper, which I am sure we all know, even if we do not know another constellation in the heavens.

If, then, we look at the heavens at a later hour than that for which the constellations are given we will find them farther westward and if our time of observation is earlier in the evening than the hour mentioned we will find them farther eastward.

In the course of a year the earth makes one trip around the sun and faces in turn all parts of the heavens. That is, it turns through an angle of 360° with respect to the heavens in a year or through an angle of 360° ÷ 12 or 30° in one month. So as a result of our revolution around the sun, which is also in a west to east direction, we see that all the constellations are gradually shifting westward at the rate of 30° a month. It is for this reason that we see different constellations in different months, and it is because of the turning of the earth on its axis that we see different constellations at different hours of the night.

If we should sit up from sunset to sunrise and watch the stars rise in the east, pass the meridian and set in the west—as the sun does by day—we should see in turn the same constellations that are to pass across the heavens in the next six months. This is because in twelve hours' time we are carried through the same angle with respect to the heavens by the earth's rotation on its axis that we are in the next six months by the motion of the earth around the sun.

Let us suppose then that the time we choose for our observation of the heavens is the last of the month while our charts are given for the first of the month. We must look then farther westward for our constellations just as we must look farther westward if we chose a later hour in the evening for our observations. Let us suppose that we choose for our time of observation half-past eight in the early part of December. On or close to the meridian we will find the constellations outlined in the charts for December. To the east of the meridian we will find the constellations that are given for January and February, and to the west of the meridian the constellations that are given for November and October. So if we are particularly ambitious or wish to become acquainted with the constellations more rapidly we may study at the same time the constellations for the preceding months now west of the meridian and the constellations for the following months now east of the meridian as well as the constellations for the month which will be due north or south or directly overhead as the case may be.

If we were able to see the stars by day as well as by night we would observe that as the days go by the sun is apparently moving continuously eastward among certain constellations. This is a result of the earth's actual motion around the sun in the same direction.

The apparent path of the sun among the stars is called the ecliptic and the belt of the heavens eight degrees wide on either side of the ecliptic is called the zodiac. The constellations that lie within this belt of the zodiac are called zodiacal constellations. The zodiac was divided by the astronomer Hipparchus, who lived 161-126 B.C., into twelve signs 30° wide, and the signs were named for the constellations lying at that time within each of these divisions. These zodiacal constellations are Aries, Taurus, Gemini, Cancer, Leo, Virgo, Libra, Scorpio, Sagittarius, Capricornus, Aquarius and Pisces. With the exception of Libra, the Scales, all of these constellations are named for people or animals and the word zodiac is derived from the Greek word meaning "of animals."

Each month the sun moves eastward 30° through one of these zodiacal constellations. In the days of Hipparchus the sun was in Aries at the beginning of spring, at the point where the ecliptic crosses the celestial equator—which lies directly above the earth's equator. This point where the ecliptic crosses the equator was then known as the First Point in Aries. The autumnal equinox was 180° distant in Aquarius and the two points were called the equinoxes because when the sun is at either equinox the day and night are equal in length all over the world. Now for certain reasons which we will not explain here the equinoctial points are not fixed in position but shift gradually westward at the rate of 1° in 70 years. It is as if the equinoxes were advancing each year to meet the sun on its return and their westward motion is therefore called "The Precession of the Equinoxes."

Since the days of Hipparchus this motion has amounted to about 30° so that the constellations no longer occupy the signs of the zodiac that bear their names.

The sun is now in Pisces instead of Aries at the beginning of spring and in Virgo instead of Aquarius at the beginning of fall.

Not only the sun but the moon and planets as well move through the zodiacal constellations. In fact a limit for the zodiac of 8° on either side of the ecliptic was chosen because it marks the extent of the excursions of the moon and planets from the ecliptic. Neither moon nor planets will be found at a greater distance than 8° on either side of the ecliptic.

For convenience in determining the positions of the heavenly bodies the astronomer assumes that they lie upon the surface of a celestial sphere that has its center at the center of the earth.

The north pole of the celestial sphere lies directly above the north pole of the earth and the south pole of the celestial sphere directly above the south pole of the earth. The celestial equator is the great circle of the celestial sphere that lies midway between its north and south poles and directly above the earth's equator. The ecliptic is also a great circle of the celestial sphere and cuts the celestial equator at an angle of 23½° in the two points 180° apart known as the equinoctial points, of which we have already spoken.

The zodiacal constellations lie nearly overhead within the tropics and can be seen to advantage all over the world except in polar regions.

For every position of the earth's surface except at the equator we have also our circumpolar constellations which are the ones that never pass below the horizon for the place of observation.

In 40° N. Latitude the Big Dipper is a circumpolar constellation for it is above the horizon at all hours of the day and night and all times of the year. If our latitude is 40° N., all stars within 40° of the north pole of the heavens are circumpolar and never set, while stars within 40° of the south pole of the heavens never rise. All other stars rise and set daily.

If we were at the north pole all stars within 90° of the north pole of the heavens would be circumpolar and would describe daily circuits of the pole parallel to the horizon remaining always above it.

If we were at the equator all stars within zero degrees of either pole would be circumpolar, that is no stars would be circumpolar, all stars rising and setting daily.

As a general rule, then, we may say that stars within an angular distance of the nearest pole of the heavens equal to the latitude never set and stars within an equal distance of the opposite pole never rise while all stars outside of these limits rise and set daily.

The beginner who attempts to make the acquaintance of the principal stars and constellations occasionally may find a bright star in a constellation that is not noted in the diagrams. In this case he has probably happened upon one of the bright planets.

It is not possible to include the planets in our diagrams for the reason that they are not fixed in position but apparently wander among the stars. The name planet is, in fact, derived from a Greek word meaning "wanderer." The stars shine by their own light but the planets shine only by reflected light from the sun. Of the seven planets in the solar system additional to our own planet earth, there are two, Uranus and Neptune that we need not consider for Neptune is not visible without the aid of a telescope and Uranus is fainter than any of the stars included in our diagrams.

Mercury will never appear except in the morning or evening twilight, when none but the very brightest stars are visible, since it never departs far from the sun. It will only be seen under certain favorable conditions, and usually it will escape our notice altogether unless we know exactly where to look for it although there are but two or three stars in the heavens that surpass it in brightness.

Venus, we will probably never mistake for any star in the heavens for it far surpasses all stars in brightness. It will always be seen in the west after sunset or in the east before sunrise and it is never seen more than three hours before or after the sun.

This leaves us but three planets, Jupiter, Saturn and Mars that we may mistake for bright stars. There is little chance that Jupiter will be thus mistaken for it also is far brighter than all of the stars except Sirius which differs greatly from Jupiter in color. Sirius is a brilliant white and Jupiter is a golden yellow. The planets do not twinkle as the stars do and this is particularly true of Jupiter which is remarkable for the quiet steadiness of its yellow light. This alone would serve to identify it.

Saturn is probably mistaken for a star oftener than any of the other planets. It moves so slowly among the stars that we would have to watch it for a number of successive evenings before we could discover that it is moving with respect to the stars. Saturn is yellowish in color and we can probably best distinguish it by the steadiness of its light. If we find in one of the zodiacal groups of stars—for the planets appear among no other constellations—a bright yellowish star where no bright star is indicated on the diagram we may be reasonably certain that we have found the planet Saturn.

Mars is the only planet that is reddish in color. Once in fifteen or seventeen years, when it is particularly near to the earth, it surpasses even Jupiter in brightness, but ordinarily it appears no more brilliant than one of the brighter stars. There are only two stars with which we are likely to confuse Mars,—Aldebaran and Antares—which are very similar to it in color, and, at times, in brightness. Moreover, both of these stars are zodiacal stars and Mars frequently passes through the constellations to which they belong. There should be no trouble about identifying Aldebaran and Antares, however, from their distinctive positions in the diagrams so that any other reddish star appearing in any of the zodiacal groups we may feel certain is the planet Mars.

In the following diagrams of the constellations the brightest and most conspicuous stars, called first-magnitude stars, are represented by white stars. These are the stars we should all be able to recognize and call by name and in every instance the name of a first-magnitude star is given on the diagram. All other stars are represented by circles, and the size of the circle is an indication of the brightness of the star.

Stars visible without the aid of a telescope are referred to usually as "naked-eye stars." They are classed as first, second, third, four, fifth or sixth magnitude stars, according to their relative brightness. A star of the first magnitude is about two and one-half times brighter than a star of the second magnitude, which in turn is two and one-half times brighter than a star of the third magnitude and so on. A first-magnitude star is, then, one hundred times brighter than a sixth magnitude star which is the faintest star that can be seen without the aid of the telescope.

This ratio between successive magnitudes continues among the telescopic stars. A star of the sixth magnitude is one hundred times brighter than a star of the eleventh magnitude which in turn is one hundred times brighter than a star of the sixteenth magnitude.

The faintest stars that can be seen visually in the greatest telescopes are of the seventeenth or eighteenth magnitude, though stars two or three magnitudes fainter can be photographed.

The faintest stars shown in the diagrams are fifth-magnitude stars and stars of this magnitude as well as stars of the fourth magnitude are only given when needed to fill out the distinctive outlines of the constellations which have been formed by connecting the principal stars in each group by dotted lines.

All stars of first, second and third magnitude are given in the diagrams without exceptions as such stars are visible to everyone on clear nights.

The constellations given in the following pages include practically all of the constellations that can be seen in 40° N. Latitude. A diagram is given for each constellation.

In this latitude it is impossible to see the constellations of the southern hemisphere that lie within 40° of the south pole of the heavens. A brief chapter with diagram treats of these constellations that are invisible in mid-latitudes of the northern hemisphere.

## CHAPTER II. JANUARY.

One of the most easily recognized constellations in the heavens is Taurus, The Bull, a zodiacal group which lies just south of the zenith in our latitudes in the early evening hours about the first of January.

Taurus is distinguished by the V-shaped group of The Hyades, which contains the bright, red, first-magnitude star Aldebaran, representing the fiery eye of the bull. It also contains the famous cluster of faint stars known as The Pleiades, lying a short distance northwest of The Hyades.

No group of stars is more universally known than The Pleiades. All tribes and nations of the world, from the remotest days of recorded history up to the present time, have sung the praises of The Pleiades. They were "The Many Little Ones" of the Babylonians, "The Seven Sisters" of the Greeks, "The Seven Brothers" of the American Indians, "The Hen and Chickens" of many nations of Europe, "The Little Eyes" of the South Sea Islanders. They were honored in the religious ceremonies of the Aztecs, and the savage tribes of Australia danced in their honor. Many early tribes of men began their year with November, the Pleiad month; and on November 17th, when The Pleiades crossed the meridian at midnight, it was said that no petition was ever presented in vain to the kings of ancient Persia.

January—Taurus

Poets of all ages have felt the charm of The Pleiades. Tennyson gives the following beautiful description of The Pleiades in Locksley Hall:
"Many a night I saw the Pleiades, rising through the mellow shade,Glitter like a swarm of fireflies tangled in a silver braid."

A well-known astronomer, not so many years ago, also felt the mysterious charm of The Pleiades and seriously expressed the belief that Alcyone, the brightest star of The Pleiades, was a central sun about which all other suns were moving. But we know that there is no foundation whatever for such a belief.

A fairly good eye, when the night is clear and dark, will make out six stars in this group arranged in the form of a small dipper. A seventh star lies close to the star at the end of the handle and is more difficult to find. It is called Pleione, and is referred to in many legends as the lost Pleiad. Persons gifted with exceptionally fine eyesight have made out as many as eleven stars in the group; and with the aid of an ordinary opera-glass, anyone can see fully one hundred stars in this cluster. Astronomers have found that The Pleiades cluster contains at least two hundred and fifty stars, all drifting slowly in the same general direction through space, and that the entire group is enveloped in a fiery, nebulous mist which is most dense around the brightest stars. It is not known whether the stars are being formed from the nebula or whether the nebula is being puffed off from the stars. The brightest star, Alcyone, is at least two hundred times more brilliant than our own sun, and all of the brighter stars in the group surpass the sun many times in brightness. It is believed that this cluster is so large that light takes many years to cross from one end of it to the other, and that it is so far from the earth that its light takes over three centuries to reach us, traveling at the rate of 186,000 miles a second.

The Hyades is a group of stars scarcely less famous than The Pleiades, and the stars in the group also form a moving cluster of enormous extent at a distance of 140 light-years from the earth.

Among the ancients, The Hyades were called the rain-stars, and the word Hyades is supposed to come from the Greek word for rain. Among the many superstitions of the past was the belief that the rising or setting of a group of stars with the sun had some special influence over human affairs. Since The Hyades set just after the sun in the showery springtime and just before sunrise in the stormy days of late fall, they were always associated with rain. In Tennyson's Ulysses we read:
"Through scudding drifts the rainy HyadesVex'd the dim sea."

The Hyades outline the forehead of Taurus, while two bright stars some distance to the northeast of the V form the tips of the horns. Only the head and forequarters of the bull are shown in the star-atlases that give the mythological groups, for, according to one legend, he is swimming through the sea and the rest of his body is submerged. According to another legend, Taurus is charging down upon Orion, The Warrior, represented by the magnificent constellation just to the southeast of Taurus, of which we shall have more to say next month.

Aldebaran is the Arabic word for "The Hindmost," and the star is so called because it follows The Pleiades across the sky. It is one of the most beautiful of all the many brilliant stars visible at this time and we might profit by following the advice of Mrs. Sigourney in The Stars:
"Go forth at nightAnd talk with Aldebaran, where he flamesIn the cold forehead of the wintry sky."

Next to Aldebaran in the V is the interesting double star Theta, which we can see as two distinct stars without a telescope.

Directly south of Taurus is Eridanus, sometimes called Fluvius Eridanus, or The River Eridanus. Starting a little to the southeast of Taurus, close to the brilliant blue-white star Rigel in Orion, it runs to the westward for a considerable distance in a long curving line of rather faint stars, bends sharply southward for a short distance, then curves backward toward the east once more, and, after running for some distance, makes another sharp curve to the southwest and disappears below the southern horizon. Its course is continued far into the southern hemisphere. Its brightest star, Achernar, is a star of the first magnitude, but it lies below the horizon in our latitudes.

January—Eridanus

Eridanus contains no star of particular interest to us. Most of the numerous stars that mark its course are of the fourth and fifth magnitude. It contains but two stars of the third magnitude, one at the beginning of its course and one close to the southwestern horizon. The beautiful constellation of Perseus lies just to the north of Taurus and should rightfully be considered among the constellations lying nearest to the meridian in January, but we give this constellation among the star groups for December because of its close association with the nearby constellations Andromeda and Pegasus in legend and story.

## CHAPTER III. FEBRUARY.

Across the meridian, due south, between eight and nine o'clock in the evening in the early part of February, lies Orion, The Warrior, generally considered to be the finest constellation in the heavens. Orion is directly overhead at the equator, and so is seen to advantage from all parts of the world except the extreme northern and southern polar regions.

A group of three faint stars outlines the head of Orion. His right shoulder is marked by the deep-red, first-magnitude star Betelgeuze (meaning armpit), and his left shoulder by the bright white star Bellatrix, The Amazon. Orion stands facing Taurus, The Bull, and brandishes in his right hand a club, outlined by a number of faint stars extending from Betelgeuze toward the northeast. The top of the club lies near the tips of the horns of Taurus. In his left hand he holds up a lion's skin, which we can trace in another curving line of faint stars to the west and northwest of Bellatrix. The brilliant, blue-white, first-magnitude star Rigel lies in the left foot, and the second-magnitude star Saiph, a little to the east of Rigel, is in the right knee. Three evenly spaced stars lying in a straight line that is exactly three degrees in length form the Belt of Orion, and from the Belt hangs the Sword of Orion, outlined by three faint stars. The central star in the Sword appears somewhat blurred and is the multiple star Theta, in the midst of the great Orion nebula, the finest object of its kind in the heavens. Entangled in the meshes of this glowing nebula are a number of brilliant suns, appearing to us as faint stars because of their great distance. The star Theta, in the heart of the nebula, is seen with a powerful telescope to consist of six stars; that is, it is a sextuple star. Even with a small telescope, four of these stars can readily be seen, arranged in the form of a small trapezium. The lowest star in the Sword is a triple star, and the entire constellation abounds in double, triple, and multiple stars.

From the central portion of the nebula extend many branches and streamers of nebulous light, and it is known that the entire constellation of Orion is enwrapped in the folds of this nebulosity, which forms a glowing, whirling mass of fiery gases in rapid rotation. This constellation is remarkable for the fact that all of its brighter stars, with the exception of the deep-red Betelgeuze, form one enormous, connected group of stars. They are all more or less associated with the great nebula and its branches, and are all extremely hot, white or bluish-white stars, known as helium stars, because the gas helium is so conspicuous in their atmospheres. The Orion stars are the hottest and brightest of all the stars.

Blazing Rigel, Bellatrix, and Saiph, marking three corners of the great quadrilateral, of which Betelgeuze marks the fourth corner, are all brilliant helium stars. So are the three stars in the Belt and the fainter stars in the Sword and the great nebula.

February—Orion

It has been estimated that the great Orion group of stars is over six hundred light-years from the earth, or about forty million times more distant than the sun. For more than six centuries the rays of light that now enter our eyes from these stars have been traveling through space with the speed of lightning. So we see Orion not as it exists today, but as it was six centuries ago. The extent of the Orion group of stars is also inconceivably great. Even the central part of the great nebula, which appears to our unaided eyes only as a somewhat fuzzy star, would extend from here to the nearest star and beyond, while our entire solar system would be the merest speck in its midst.

Betelgeuze, the red star that marks the right shoulder of Orion, is, as we have said, not a member of the Orion group. It has been estimated that it is about two hundred light-years from the earth, or only about one-third as far away as the other stars of the constellation.

Betelgeuze very recently has attracted universal attention, and will probably be considered an object of historic interest in the future, because it is the first star to have its diameter measured with the new Michelson interferometer, which is now being used so successfully to measure the diameters of the largest stars. The truly sensational discovery has been made that Betelgeuze is a supergiant of the universe, with a diameter of about 275,000,000 miles. Our own sun, which is known as a "dwarf" star, has a diameter of 864,000 miles. That is, Betelgeuze would make about thirty million suns the size of our own. If placed at the center of the solar system, it would fill all of the space within the orbit of Mars; and the planets Mercury, Venus, and the Earth would lie far beneath its surface. Measurements of the diameters of other giant stars which are now being made with the interferometer give results quite as startling as have been obtained in the case of Betelgeuze; and it has been found that several of these stars may even exceed Betelgeuze in size. Such a star is Antares, the fiery-red star in the heart of Scorpio, which is such a conspicuous object in the summer evening skies. All these huge stars are deep red in color, and some of them vary irregularly in brightness. Betelgeuze is one of the stars that changes in brightness in a peculiar manner from time to time. When shining with its greatest brilliancy it is a brighter object than the nearby star Aldebaran, in Taurus; but a few months or a year later it may lose so much of its light as to be decidedly inferior to Aldebaran. We may note for ourselves this remarkable change in the brightness of Betelgeuze by comparing the two stars from time to time.

Directly south of Orion lies the small constellation of Lepus, The Hare, which is made up of third-magnitude and fourth-magnitude stars. The four brighter stars are arranged in the form of a small, but distinct, quadrilateral, or four-sided figure, which may be easily seen in our latitudes. The small constellation of Columba, The Dove, which lies just south of Lepus, is so close to the horizon that it can not be seen to advantage in the mid-latitudes of the northern hemisphere. Neither Lepus nor Columba contain any object of unusual interest.

February—Auriga

Due north of Orion, and lying in the zenith at this time, is Auriga, The Charioteer, represented, strange to say, with Capella, a goat, in his arms. The beautiful first-magnitude star Capella, golden-yellow in color, serves us in identifying the constellation. Close at hand are The Kids, represented by a group of three faint stars. Capella is one of the most brilliant stars of the northern hemisphere. It is almost exactly equal in brightness to Arcturus and Vega, stars conspicuous in the summer months, and it is a shade brighter than magnificent blue-white Rigel in Orion. Capella is about fifty light-years distant from the earth and is fully two hundred times more brilliant than our own sun. At the distance of Capella, the sun would appear to be considerably fainter than any one of the three stars in the nearby group of The Kids.

Capella is attended by a companion star so close to its brilliant ruler that it can not be seen as a separate star save with the aid of the most powerful telescopes. Its distance from Capella has been very accurately measured, however, by means of the interferometer, which is giving us the measurements of the diameters of the giant stars. It is known that this companion sun is closer to Capella than our planet earth is to the sun.

At no time of the year shall we find near the meridian so many brilliant and beautiful stars as appear in the month of February at this time in the evening. In addition to Capella, which is one of the three most brilliant stars in the northern hemisphere of the heavens, we have, in Orion alone, two stars of the first magnitude, Betelgeuze and Rigel, and five stars of the second magnitude, Bellatrix and Saiph and the three stars in the Belt. In addition, we have not far distant in the western sky, fiery Aldebaran in Taurus, and close on the heel of Orion in the east, Sirius, the brightest star in the heavens, in the constellation of Canis Major, The Greater Dog, as well as the first-magnitude star Procyon in Canis Minor, The Lesser

Dog. Of these two groups we shall have more to say under the constellations for March.

## CHAPTER IV. MARCH.

To the southeast of Orion and almost due south at eight o'clock in the evening on the first of March lies the constellation of Canis Major, The Greater Dog, containing Sirius, the Dog-star, which far surpasses all other stars in the heavens in brilliancy.

Sirius lies almost in line with the three stars that form the Belt of Orion. We shall not have the slightest difficulty in recognizing it, owing to its surpassing brilliancy as well as to the fact that it follows so closely upon the heels of Orion.

Sirius is the Greek for "scorching" or "sparkling," and the ancients attributed the scorching heat of summer to the fact that Sirius then rose with the sun. The torrid days of midsummer they called the "dog-days" for this reason, and we have retained the expression to the present time. Since Sirius was always associated with the discomforts of the torrid season, it did not have an enviable reputation among the Greeks. We find in Pope's translation of the Iliad this reference to Sirius: "Terrific glory! for his burning breathTaints the red air with fever, plagues, and death."

In Egypt, however, many temples were dedicated to the worship of Sirius, for the reason that some five thousand years ago it rose with the sun at the time of the summer solstice, which marks the beginning of summer, and heralded the approaching inundation of the Nile, which was an occasion for great rejoicing among the Egyptians. It was, therefore, called the Nile Star and regarded by them with the greatest reverence.

Sirius is an intensely white hydrogen star; but owing to its great brilliancy and to the fact that it does not attain a great height above the horizon in our latitudes, its rays are greatly refracted or broken up by the atmosphere, which is most dense near the horizon, and as a result, it twinkles or scintillates more noticeably than other stars and flashes the spectrum colors—chiefly red and green—like a true "diamond in the sky"—a magnificent object in the telescope.

Sirius is one of our nearest neighbors among the stars. Only two stars are known to be nearer to the solar system. Yet its light takes about eight and a half years to flash with lightning speed across the great intervening chasm. It is attended also by a very faint star that is so lost in the rays of its brilliant companion that it can only be found with the aid of a powerful telescope. The two stars are separated by a distance of 1,800,000,000 miles; that is they are about as far apart as Neptune and the sun. They swing slowly and majestically about a common center, called their center of gravity, in a period of about forty-nine years. So faint is the companion of Sirius that it is estimated that twenty thousand such stars would be needed to give forth as much light as Sirius. The two stars together, Sirius and its companion, give forth twenty-six times as much light as our own sun. They weigh only about three times as much, however. The companion of Sirius, in spite of its extreme faintness, weighs fully half as much as the brilliant star.

March—Canis Major

There are a number of bright stars in the constellation of Canis Major. A fairly bright star a little to the west of Sirius marks the uplifted paw of the dog, and to the southeast, in the tail and hind quarters, are several conspicuous stars of the second magnitude.

A little to the east and much farther to the north, we find Canis Minor, The Lesser Dog, containing the beautiful first-magnitude star Procyon, "Precursor of the Dog"—that is, of Sirius. Since Procyon is so much farther north than Sirius and very little to the east, we see its brilliant rays in the eastern sky some time before Sirius appears above the southeastern horizon, hence its name. Long after Sirius has disappeared from view beneath the western horizon in the late spring, Procyon may still be seen low in the western sky. Procyon, also is one of our nearer neighbors among the stars, being only about ten light-years distant from the solar system. Like Sirius, it is a double star with a much fainter companion, that by its attraction sways the motion of Procyon to such an extent that we should know of its existence, even if it were not visible, by the disturbances it produces in the motion of Procyon. The period of revolution of Procyon and its companion about a common center is about forty years, and the two stars combined weigh about a third more than our own sun and give forth six times as much light. Canis Minor contains only one other bright star, Beta, a short distance to the northwest of Procyon. Originally, the name Procyon was given to the entire constellation, but it was later used only with reference to the one star. Procyon, Sirius, and Betelgeuze in Orion form a huge equal-sided triangle that lies across the meridian at this time and is a most conspicuous configuration in the evening sky.

March—Gemini and Canis Minor

Directly south of the zenith we find Gemini, The Twins, one of the zodiacal constellations. It is in Gemini that the sun is to be found at the beginning of summer. The two bright stars Castor and Pollux mark the heads of the twins, and the two stars in the opposite corners of the four-sided figure shown in the chart mark their feet.

Castor and Pollux, according to the legend, were the twin brothers of Helen of Troy who went on the Argonautic expedition. When a storm overtook the vessel on its return voyage, Orpheus invoked the aid of Apollo, who caused two stars to shine above the heads of the twins, and the storm immediately ceased. It was for this reason that Castor and Pollux became the special deities of seamen, and it was customary to place their effigies upon the prows of vessels. The "By Jimini!" of today is but a corruption of the "By Gemini!" heard so frequently among the sailors of the ancient world.

The astronomical name for Castor, the fainter star, is Alpha Geminorum, meaning Alpha of Gemini. As it was customary to call the brightest star in a constellation by the first letter in the Greek alphabet, it is believed that Castor has decreased considerably in brightness since the days of the ancients, for it is now decidedly inferior to Pollux in brightness, which is called Beta Geminorum. Of the two stars, Castor is the more interesting because it is a double star that is readily separated into two stars with the aid of a small telescope. The two principal stars are known to be, in turn, extremely close double stars revolving almost in contact in periods of a few days. Where we see but one star with the unaided eye, there is, then a system of four suns, the two close pairs revolving slowly about a common center of gravity in a period of several centuries and at a great distance apart.

The star Pollux, which we can easily distinguish by its superior brightness, is the more southerly of the twin stars and lies due north of Procyon and about as far from Procyon as Procyon is from Sirius.

The appearance of Gemini on the meridian in the early evening and of the huge triangle, with its corners marked by the brilliants, Procyon, Sirius, and Betelgeuze, due south, with "Great Orion sloping slowly to the west," is as truly a sign of approaching spring as the gradual lengthening of the days, the appearance of crocuses and daffodils, and the first robin. It is only a few weeks later—

## CHAPTER V. APRIL.

In the early evening hours of April the western sky is still adorned with the brilliant jewels with which we became familiar on the clear frosty evenings of winter. Orion is now sinking fast to his rest beneath the western horizon. Beautiful, golden Capella in Auriga glows in the northwest. Sirius sparkles and scintillates, a magnificent diamond of the sky, just above the southwestern horizon, while Procyon in Canis Minor, The Lesser Dog, and Castor and Pollux, The Twins, in the constellation of Gemini, are still high in the western part of the heavens.

In the northeast and east may be seen the constellations that will be close to the meridian at this time next month. Ursa Major, The Greater Bear, with its familiar Big Dipper, is now in a favorable position for observation. The Sickle in Leo is high in the eastern sky, and Spica, the brilliant white diamond of the evening skies of spring, is low in the southeast in Virgo.

Near the meridian this month we find between Auriga and Ursa Major, and east of Gemini, the inconspicuous constellation of Lynx, which contains not a single bright star and is a modern constellation devised simply to fill the otherwise vacant space in circumpolar regions between Ursa Major and Auriga.

April—Cancer

Just south of the zenith at this time, and lying between Gemini and Leo, is Cancer, The Crab, the most inconspicuous of all the zodiacal constellations. There are no bright stars in this group, and there is also nothing distinctive about the grouping of its faint stars, though we can readily find it, from its position between the two neighboring constellations of Gemini and Leo by reference to the chart.

In the position indicated there we will see on clear evenings a faint, nebulous cloud of light, which is known as Praesepe, The Beehive, or as The Manger, the two faint stars flanking it on either side being called Aselli, The Asses. This faint cloud can be easily resolved by an opera-glass into a coarse cluster of stars that lie just beyond the range of the unaided human vision.

To the ancients, Praesepe served as an indicator of weather conditions, and Aratus, an ancient astronomer, wrote of this cluster:
"A murky manger, with both starsShining unaltered, is a sign of rain.If while the northern ass is dimmedBy vaporous shroud, he of the south gleam radiant,Expect a south wind; the vaporous shroud and radianceExchanging stars, harbinger Boreas."

This was not entirely a matter of superstition, as we might possibly imagine, for the dimness of the cluster is simply an indication that vapor is gathering and condensing in the atmosphere, just as a ring around the moon is an indication of the same gathering and condensation of vapor that precedes a storm.

Some centuries ago the sun reached its greatest distance north of the equator—which occurs each year at the beginning of summer—at the time when it was passing through the constellation of Cancer. Our tropic of Cancer, which marks the northern limit of the torrid zone, received its name from this fact. At the time when the sun reaches the point farthest north, its height above the horizon changes very little from day to day, and for a short time it appears to be slowly crawling sideways through the heavens, as a crab walks, and for this reason, possibly, the constellation was called Cancer, The Crab. At the present time the "Precession of the Equinoxes," or westward shifting of the vernal equinox—the point where the sun crosses the equator going north in the spring—brings the sun, when it is farthest north, in Gemini instead of in Cancer. At the present time, then it would be more accurate to speak of the tropic of Gemini, though this in turn would be inaccurate after a lapse of centuries, as the sun passed into another constellation at the beginning of summer. The tropic of Capricorn, which marks the farthest southern excursions of the sun in its yearly circuit of the heavens, should also more appropriately be called the tropic of Sagittarius, as the sun is now in Sagittarius instead of Capricornus at the time when it is farthest south, though the point is slowly shifting westward into Scorpio.

Mythology tells us that Cancer was sent by Juno to distract Hercules by pinching his toes while he was contending with the many-headed serpent in the Lernean swamp. Hercules, the legend says, crushed the crab with a single blow, and Juno by way of reward placed it in the heavens.

In Cancer, according to the belief of the Chaldeans, was located the "gate of men," by which souls descended into human bodies, while in Capricornus was the "gate of the gods," through which the freed souls of men returned to heaven.

## April—Hydra

Hydra, the many-headed serpent with which Hercules contended, is represented by a constellation of great length. It extends from a point just south of Cancer, where a group of faint stars marks the heads, to the south and southeast in a long line of faint stars. It passes in its course just south of Crater and Corvus, the two small star-groups below Leo (see constellations for May), which are sometimes called its riders, and it also stretches below the entire length of the long, straggling constellation of Virgo. At this time we can trace it only to the point where it disappears below the horizon in the southeast. It contains but one bright star, Alphard, or Cor Hydrae as it is also called, standing quite alone and almost due south at this time. Hydra, as well as Lynx and Cancer, contains no noteworthy or remarkable object and consists chiefly of faint stars. Alphard is, in fact, the only bright star that we have in the constellations for this month. It chances that these three inconspicuous star-groups, Lynx, Crater, and Hydra, lie nearest to the meridian at this time, separating the brilliant groups of winter from those of the summer months.

April—Lynx

## CHAPTER VI. MAY.

Ursa Major, the Great Bear, and Ursa Minor, the Lesser Bear, or, as they are more familiarly called, the Big Dipper and the Little Dipper, are the best known of all the constellations visible in northern latitudes. They are called circumpolar constellations, which means "around the pole." For those who live north of 40° N. Lat. they never set, but can be seen at all hours of the night and at all times of the year. In fall and winter evenings Ursa Major lies below the pole and near the horizon, and so is usually hidden more or less from view by trees or buildings. It is during the early evening hours of late spring and summer that this constellation is seen to the best advantage high in the sky above the pole. If one looks due north at the time mentioned, it will be impossible to miss either of these constellations.

To complete the outline of the Great Bear, it is necessary to include faint stars to the east, which form the head of the Bear, and other faint stars to the south, which form the feet, but these are all inconspicuous and of little general interest.

The two stars in the bowl of the Big Dipper through which an arrow is drawn in the chart, are called the Pointers, because an imaginary line drawn through these two stars and continued a distance about equal to the length of the Big Dipper, brings us to the star Polaris, or the North Star, at the end of the handle of the Little Dipper, which is very close to the north pole of the heavens, the direction in which the earth's axis points. The pole lies on the line connecting the star at the bend in the handle of the Big Dipper with Polaris, and is only one degree distant from the pole-star.

May—Ursa Major and Ursa Minor

The distance between the Pointers is five degrees of arc, and the distance from the more northerly of these two stars to Polaris is nearly thirty degrees. We may find it useful to remember this in estimating distances between objects in the heavens, which are always given in angular measure.

A small two and one-half inch telescope will separate Polaris into two stars eighteen seconds of arc apart. The companion star is a faint white star of the ninth magnitude.

Twenty years or so ago it was discovered with the aid of the spectroscope that the brighter of the two stars was also a double star, but the two stars were so close together that they could not be seen as separate stars in any telescope. Later it was found that the brighter star was in reality triple, that is, it consists of three suns close together. The faint white companion star formed with these three suns a system of four suns revolving about a common center of gravity. Still more recently it has been discovered that the brightest of these four suns varies regularly in brightness in a period of a little less than four days. It belongs to the important class of stars known as Cepheid variable stars, whose changes of light, it is believed, are produced by some periodic form of disturbance taking place within the stars themselves.

With one exception, Polaris is the nearest to the earth of all these Cepheid variable stars, which are in most instances at very great distances from the solar system. The latest measurements of the distance of Polaris show that its light takes about two centuries to travel to the earth, or, in other words, that it is distant two hundred light-years.

Like all Cepheid variables, Polaris is a giant star. It gives forth about five hundred and twenty-five times as much light as our own sun. If Polaris and the sun were placed side by side at a distance of thirty-three light-years, the sun would appear as a star of the fifth magnitude, just well within the range of visibility of the human eye, while Polaris would outshine Sirius, the brightest star in the heavens.

As a practical aid to navigators, Polaris is unsurpassed in importance by any star of the northern hemisphere of the heavens. At the equator the pole-star lies in the horizon; at the north pole of the earth it is in the zenith or directly overhead. Its altitude or height above the horizon is always equal to the latitude of the place of observation. As we travel northward from the equator toward the pole we see Polaris rise higher and higher in the sky. In New York the elevation of Polaris above the horizon is forty degrees, which is the latitude of the city.

The Pointers indicate the direction of Polaris and the true north, while the height of Polaris above the horizon tells us our latitude. These kindly stars direct us by night when we are uncertain of our bearings, whether we travel by land or sea or air. They are the friends and aids of explorers, navigators and aviators, who often turn to them for guidance.

Bryant writes thus beautifully of Polaris in his Hymn to the North Star:
Constellations come and climb the heavens, and go.Star of the Pole! and thou dost see them set.Alone in thy cold skies,Thou keep'st thy old unmoving station yet,Nor join'st the dances of that glittering train,Nor dipp'st thy virgin orb in the blue western main.

On thy unaltering blazeThe half wrecked mariner, his compass lost,Fixes his steady gaze,And steers, undoubting, to the friendly coast;And they who stray in perilous wastes by night,Are glad when thou dost shine to guide their footsteps right.

The star at the bend in the handle of the Big Dipper, called Mizar, is of special interest. If one has good eyesight, he will see close to it a faint star. This is Alcor, which is Arabic for The Test. The two stars are also called the Horse and the Rider.

Mizar forms with Alcor what is known as a wide double star. It is, in fact, the widest of all double stars. Many stars in the heavens that appear single to us are separated by the telescope into double or triple or multiple stars. They consist of two or more suns revolving about a common center, known as their center of gravity. Sometimes the suns are so close together that even the most powerful telescope will not separate them. Then a most wonderful little instrument, called the spectroscope, steps in and analyzes the light of the stars and shows which are double and which are single. A star shown to be double by the spectroscope, but not by the telescope, is called a spectroscopic binary star.

Mizar is of historic interest, as being the first double star to be detected with the aid of the telescope. A very small telescope will split Mizar up into two stars. The brighter of the two is a spectroscopic binary star beside, so that it really consists of two suns instead of one, with the distance between the two so small that even the telescope cannot separate them. About this system of three suns which we know as the star Mizar, the faint star Alcor revolves at a distance equal to sixteen thousand times the distance of the earth from the sun.

May—Leo

If we follow the imaginary line drawn through the Pointers in a southerly direction about forty-five degrees, we come to Leo, The Lion, one of the zodiacal constellations. There should be no difficulty in finding the constellation Leo, as its peculiar sickle-shaped group of bright stars makes it distinctive from all other constellations. At the time we have mentioned, that is, the early evening hours, it will lie a little to the southwest of the zenith. Leo is one of the finest constellations and is always associated with the spring months because it is then high in the sky in the evening. Regulus is the beautiful white star which marks the handle of The Sickle, and the heart of Leo; and Denebola is the second-magnitude star in the tail of Leo.

## May—Corvus and Crater

Due south of Denebola, about thirty degrees, we find the small star-group known as Crater, The Cup, which is composed of rather faint and inconspicuous stars. Just east of Crater is the group known as Corvus, The Crow, which forms a very characteristic little four-sided figure of stars differing very little from one another in brightness. These two star groups lie far to the south in our latitudes; but if we lived twenty degrees south of the equator, we would find them nearly overhead, at this time of the year. Just south of Corvus and Crater we find Hydra, one of the constellations for April which extends beneath these constellations and also beneath Virgo, one of the June constellations.

# CHAPTER VII. JUNE.

The star-groups that occupy the center of the celestial stage in mid-latitudes of the northern hemisphere during the early evening hours of June are Boötes, often called The Hunter, (although the word means Herdsman or Shouter), which will be found overhead at this time; Virgo, The Maiden, largest of the zodiacal constellations, lying nearly due south; Canes Venatici, The Hunting Dogs; Corona Borealis, The Northern Crown, and Coma Berenices.

The gorgeous orange-hued Arcturus in Boötes and the beautiful bluish-white Spica in Virgo, like a diamond in its sparkling radiance, form with Denebola in Leo, which we identified in May, a huge equal-sided triangle that is always associated with the spring and early summer months.

To the west of Boötes, below the handle of the Big Dipper, is a region where there are few conspicuous stars. Here will be found Canes Venatici (The Hunting Dogs with which Boötes is supposed to be pursuing the Great Bear around the north pole), and, further south, Coma Berenices (Bernice's Hair).

The brighter of the two Hunting Dogs, which is also the brightest star in the entire region covered by these two constellations, appears as a beautiful blue-and-yellow double star in the telescope. It was named Cor Caroli (Heart of Charles) by the astronomer Halley in honor of Charles II of England, at the suggestion of the court physician, who imagined it shone more brightly than usual the night before the return of Charles to London. Of more interest to astronomers is the magnificent spiral nebula in this constellation, known as the "Whirlpool Nebula," appearing as a faint, luminous patch in the sky, of which many photographs have been taken with the great telescopes. This entire region, from Canes Venatici to Virgo, abounds in faint spiral nebulæ that for some reason not yet understood by astronomers are crowded together in this part of the heavens where stars are comparatively few. It is believed that there are between five hundred thousand and a million of these spiral nebulæ in the entire heavens, and the problem of their nature and origin and distance is one that the astronomers are very anxious to solve. Many wonderful facts are now being learned concerning these faint nebulous wisps of light which, with a few exceptions, are observable only with great telescopes. They reveal their spiral structure more clearly to the photographic plate than to the human eye, and some magnificent photographs of them have been taken with powerful telescopes.

Coma Berenices, south of Canes Venatici and southwest of Boötes, is a constellation that consists of a great number of stars closely crowded together, and just barely visible to the unaided eye. As a result, it has the appearance of filmy threads of light, which doubtless suggested its name to the imaginative ancients, who loved to fill the heavens with fanciful creations associated with their myths and legends. These stars form a moving cluster of stars estimated to be at a distance of about 270 light-years from the solar systems.

June—Boötes, Canes Venatici and Coma Berenicis

This region, so lacking in interesting objects for the naked-eye observer, is a mine of riches to the fortunate possessors of telescopes; and the great telescopes of the world are frequently pointed in this direction, exploring the mysteries of space that abound here.

Just to the east of Boötes is the exquisite little circlet of stars known as Corona Borealis, the Northern Crown. It consists of six stars arranged in a nearly perfect semicircle, and one will have no difficulty in recognizing it. Its brightest star, Alpha, known also by the name of Alphacca, is a star of the second magnitude.

Boötes is one of the largest and finest of the northern constellations. It can be easily distinguished by its peculiar kite-shaped grouping of stars or by the conspicuous pentagon (five-sided figure) of stars which it contains. The most southerly star in this pentagon is known as Epsilon Boötes and is one of the finest double stars in the heavens. The two stars of which it consists are respectively orange and greenish-blue in color.

By far the finest object in Boötes, however, is the magnificent Arcturus, which is the brightest star in the northern hemisphere of the heavens. This star will be conspicuous in the evening hours throughout the summer months, as will also the less brilliant Spica in Virgo.

Some recent measurements show that Arcturus is one of our nearer neighbors among the stars. Its distance is now estimated to be about twenty-one light-years. That is, a ray of light from this star takes twenty-one years to reach the earth, traveling at the rate of one hundred and eighty-six thousand miles per second. It would seem as if we should hardly speak of Arcturus, twenty-one light-years away, as a near neighbor, yet there are millions of stars that are far more distant from the earth, and very few that are nearer to us than Arcturus.

The brightness of Arcturus is estimated to be about forty times that of the sun. That is, if the two bodies were side by side, Arcturus would give forth forty times as much light and heat as the sun.

Arcturus is also one of the most rapidly moving stars in the heavens. In the past sixteen centuries it has traveled so far as to have changed its position among the other stars by as much as the apparent width of the moon. Most of the stars, in spite of their motions through the heavens in various directions, appear today in the same relative positions in which they were several thousand years ago. It is for this reason that the constellations of the Egyptians and of the Greeks and Romans are the same constellations that we see in the heavens today. Were all the stars as rapidly moving as Arcturus, the distinctive forms of the constellations would be preserved for only a very few centuries.

June—Virgo

Virgo, which lies south and southwest of Boötes, is a large, straggling constellation, consisting of a Y-shaped configuration of rather inconspicuous stars. It lies in the path of our sun, moon and planets, and so is one of the zodiacal constellations. The cross in the diagram indicates the present position of the autumnal equinox, the point where the sun crosses the equator going south, and the position the sun now occupies at the beginning of fall.

Spica, the brightest star in Virgo, is a bluish-white, first-magnitude star, standing very much alone in the sky. In fact, the Arabs referred to this star as "The Solitary One." Its distance from the earth is not known, but must be very great as it cannot be found by the usual methods. The spectroscope shows that it consists of two suns very close together, revolving about a common center in a period of only four days.

Within the branches of the "Y" in Virgo, and just to the north of it, is the wonderful nebulous region of this constellation, but it takes a powerful telescope to show the faint spiral nebulæ that exist here in great profusion. All of these spirals are receding from the plane of the Milky Way with enormous velocities. The spiral nebulæ are, in fact, the most rapidly moving objects in the heavens.

## CHAPTER VIII. JULY.

Due east of the little circlet of stars known as Corona Borealis, and almost directly overhead in our latitude (40° N.) about nine o'clock in the evening in the early part of July, is the large constellation of Hercules, named for the famous hero of Grecian mythology. There are no stars of great brilliancy in this group, but it contains a large number of fairly bright stars arranged in the form outlined in the chart. The hero is standing with his head, marked by the star Alpha Herculis, toward the south, and his foot resting on the head of Draco, The Dragon, a far-northern constellation with which we become acquainted in August.

July—Hercules

Alpha Herculis, the best known star in this constellation, is of unusual interest. Not only is it a most beautiful double star, the brighter of the two stars of which it is composed being orange, and the fainter greenish-blue, but it is also a star that changes in brightness irregularly. Both the orange and the blue star share in this change of brightness. There are a number of stars in the heavens that vary in brightness, some in very regular periods, and others, like Alpha Herculis, irregularly. These latter stars are nearly always deep orange or reddish in color. One may note this variation in the brightness of Alpha Herculis by comparing it from time to time with some nearby star that does not vary in brightness.

The constellation of Hercules is a very rich field for the possessor of even a small telescope. Here are to be found beautifully colored double stars in profusion, and, in addition, two remarkable clusters of stars. The brighter of the two is known as the Great Hercules Cluster. Its position is shown on the chart, and, under favorable conditions—that is, on a clear, dark night, when there is no moonlight—it may be seen without the aid of a telescope as a small, faint patch of light. One would never suspect from such a view what a wonderful object this cluster becomes when seen with the aid of a powerful telescope. Photographs taken with the great telescopes show this faint wisp of light as a magnificent assemblage of thousands of stars, each a sun many times more brilliant than our own sun. The crowded appearance of the stars in the cluster is due partly to the fact that it is very distant from the earth, though neighboring stars in the cluster are indeed much nearer to one another than are the stars in the vicinity of our solar system. It has been found that this cluster is so far away that its light takes over thirty-six thousand years to reach the earth. At the distance of this cluster, a sun equal in brightness to our own sun would be so faint that the most powerful telescope in the world would not show it.

So we know that the stars that are visible in the Hercules cluster are far more brilliant than our sun. A fair-sized telescope will show about four thousand stars in this cluster, but the greatest telescopes show over one hundred thousand in it, and there are without doubt many more too faint to be seen at all. The Hercules cluster is called a globular star-cluster, because the stars in it are arranged nearly in the form of a sphere. There are in the heavens about ninety such clusters whose distances have been found, and they are among the most distant of all objects. Most of them are very faint, and a few are over two hundred thousand light-years distant from the earth. The Hercules cluster is one of the nearest and is the most noted of all of these globular clusters. It is considered to be one of the finest objects in the heavens. The other cluster in Hercules is also very fine, but not to be compared with this one.

## July—Ophiuchus and Serpens

Just to the south of Hercules are two constellations, Ophiuchus, The Serpent-Bearer, and Serpens, The Serpent, which are so intermingled that it is difficult to distinguish them. There are in these two constellations, as in Hercules, no stars of unusual brilliancy, but a large number of fairly bright stars. The brightest star in Ophiuchus is known as Alpha Ophiuchi and it marks the head of the Serpent-Bearer. The two stars, Alpha Ophiuchi and Alpha Herculis, are close together, being separated by a distance about equal to that between the Pointers of the Big Dipper. Alpha Ophiuchi is the brighter of the two, and it is farther east.

Ophiuchus, according to one legend, was once a physician on earth, and was so successful as a healer that he could raise the dead. Pluto, the god of the lower world, became alarmed for fear his kingdom would become depopulated, and persuaded Jupiter to remove Ophiuchus to a heavenly abode, where he would be less troublesome. The serpent is supposed to be a symbol of his healing powers. The head of Serpens is marked by a group of faint stars just south of Corona Borealis and southwest of Hercules. From here a line of fairly bright stars marks the course of Serpens southward to the hand of Ophiuchus. Two stars close together and nearly equal in brightness mark the hand with which the hero grasps the body of the serpent. The other hand is marked by an equally bright single star some distance to the eastward where the two constellations again meet. Ophiuchus is thus represented as holding the serpent with both hands. It is not an easy matter to make out the outlines of these straggling groups, but there are in them several pairs of stars nearly equal in brightness and about as evenly spaced as the two stars in the one hand of Ophiuchus, and these, as well as the diagram, will be of aid in tracing the two groups.

Just south of Serpens and Ophiuchus lies one of the most beautiful and easily recognized constellations in the heavens. This is the constellation of Scorpio, The Scorpion, which will be found not far above the southern horizon at this time. The small constellation of Libra, The Scales, which lies just to the northwest of Scorpio, was at one time a part of this constellation and represented the creature's claws, but some centuries ago its name was changed to Libra. Both Scorpio and Libra are numbered among the twelve zodiacal constellations—that is, they lie along the ecliptic, or apparent yearly path of the sun among the stars. Scorpio is the most brilliant and interesting of all the zodiacal groups. The heart of the Scorpion is marked by the magnificent first-magnitude star Antares, which is of a deep reddish color. The name signifies Rival of Ares (Mars). It is so called because it is the one star in the heavens that most closely resembles Mars, and it might be mistaken for the ruddy planet if one were not familiar with the constellations. At times, when Mars is at a considerable distance from the earth, it is almost equal in brightness and general appearance to this glowing red star in the heart of the Scorpion. In its trips around the sun, Mars passes occasionally very close to Antares, and the two then present a very striking appearance.

July—Libra and Scorpio

With a telescope of medium size, one will find an exquisite little green companion-star close to Antares. The little companion is so close to Antares that it is difficult to find it in the glare of light from its more brilliant neighbor. Antares is one of the giant stars of the universe. In fact it is, so far as we know, the greatest of all the giants. Its diameter is more than five hundred times that of our own sun and nearly twice that of the giant star Betelgeuze in Orion. If placed at the center of the solar system its surface would lie far beyond the orbit of Mars.

Both Ophiuchus and Scorpio are crossed by the Milky Way, that broad belt of numberless faint stars that encircles the heavens. Some of the most wonderful and beautiful regions of the Milky Way are to be found in these two constellations.

At various times in the past, there have suddenly flashed forth brilliant stars in the Milky Way which are known as "temporary stars," or "novæ." These outbursts signify that some celestial catastrophe has taken place, the nature or cause of which is not clearly understood. Some of the most brilliant of these outbursts have occurred in these two constellations. The life of a nova is very short, a matter of a few months, and it rapidly sinks into oblivion, so nothing is to be seen of some of the most brilliant of all these stars that have appeared in this region in the past. A few are still faintly visible in large telescopes.

## CHAPTER IX. AUGUST.

It was one of the twelve labors of Hercules, the hero of Grecian mythology, to vanquish the dragon that guarded the golden apples in the garden of the Hesperides. Among the constellations for July we found the large group of stars that represents the hero himself, and this month we find just to the north of Hercules the head of Draco, The Dragon. The foot of the hero rests upon the dragon's head, which is outlined by a group of four fairly bright stars forming a quadrilateral or four-sided figure. The brightest star in this group passes in its daily circuit of the pole almost through the zenith of London. That is, as it crosses the meridian of London, it is almost exactly overhead. From the head of Draco, the creature's body can be traced in a long line of stars curving first eastward, then northward, toward the pole-star to a point above Hercules, where it bends sharply westward. The body of the monster lies chiefly between its head and the bowl of the Little Dipper. The tail extends in a long line of faint stars midway between the two Dippers, or the constellations of Ursa Major and Ursa Minor, the tip of the tail lying on the line connecting the Pointers of the Big Dipper with the pole-star Polaris.

Draco, as well as Ursa Major and Ursa Minor, is a circumpolar constellation in our latitude; that is, it makes its circuit of the pole without at any time dipping below the horizon in latitudes north of 40°. It is, therefore, visible at all hours of the night in mid-latitudes of the northern hemisphere, but is seen to the best advantage during the early evening hours in the summer months. There are no remarkable stars in this constellation with the exception of Alpha, which lies halfway between the bowl of the Little Dipper and Mizar, the star at the bend in the handle of the Big Dipper.

August—Draco and Lyra

About four thousand seven hundred years ago, this star was the pole-star—lying even nearer to what was then the north pole of the heavens than Polaris does to the present position of the pole. The sun and moon exert a pull on the bulging equatorial regions of the earth, which tends to draw the plane of the earth's equator down into the plane of the ecliptic. This causes the "Precession of the Equinoxes" and at the same time a slow revolution of the earth's axis of rotation about the pole of the ecliptic. The north pole of the heavens as a result describes a circle about the pole of the ecliptic of radius 23½° in a period of 25,800 years.

Each bright star that lies near the circumference of this circle becomes in turn the pole-star sometime within this period. The star Alpha, in Draco, had its turn at being pole-star some forty-seven centuries ago. Polaris is now a little over a degree from the north pole of the heavens. During the next two centuries it will continue to approach the pole until it comes within a quarter of a degree of it, when its distance from the pole will begin to increase again. About twelve thousand years hence the magnificent Vega, whose acquaintance we will now make, will be the most brilliant and beautiful of all pole-stars.

Vega (Arabic for "Falling Eagle") is the resplendent, bluish-white, first-magnitude star that lies in the constellation of Lyra, The Lyre or Harp, a small, but important, constellation just east of Hercules and a little to the southeast of the head of Draco. Vega is almost exactly equal in brightness to Arcturus, the orange-colored star in Boötes, now lying west of the meridian in the early evening hours. It is also a near neighbor of the solar system, its light taking something like forty years to travel to the earth. Vega is carried nearly through the zenith of Washington and all places in the same latitude by the apparent daily rotation of the heavens. It is a star that we have no difficulty in recognizing, owing to the presence of two nearby stars that form, with it, a small equal-sided triangle with sides only two degrees in extent. If our own sun were at the distance of Vega, it would not appear as bright as one of these faint stars, so much more brilliant is this magnificent sun than our own. The two faint stars that follow so closely after Vega and form the little triangle with it are also of particular interest. Epsilon Lyræ, which is the northern one of these two stars, may be used as a test of keen eyesight. It is the finest example in the heavens of a quadruple star—that is, "a double-double star." A keen eye can just separate this star into two without a telescope, and with the aid of a telescope, each of the two splits up into two stars, making four stars in place of the one visible to the average eye. Zeta, the other of the two stars that form the little triangle with Vega, is also a fine double star. The star that lies almost in a straight line with Epsilon and Zeta and a short distance to the south of them is a very interesting variable star known as Beta Lyræ. Its brightness changes very considerably in a period of twelve days and twenty-two hours. This change of brightness is due to the presence of a companion star. The two stars are in mutual revolution, and their motion is viewed at such an angle from the earth that, in each revolution, one star is eclipsed by the other, producing a variation in the amount of light that reaches our eyes. By comparing this star from day to day with the star just a short distance to the southeast of it, which does not vary in brightness, we can observe for ourselves this change in the light of Beta Lyræ. There are a number of stars in the heavens that vary in brightness in the same manner as Beta Lyræ, and they are called eclipsing-variable stars.

On the line connecting Beta Lyræ with the star southeast of it and one-third of the distance from Beta to this star, lies the noted Ring Nebula in Lyra, which is a beautiful object even in a small telescope. It consists of a ring of luminous gas surrounding a central star. The star shines with a brilliant, bluish-white light and is visible only in powerful telescopes though it is easily photographed since it gives forth rays to which the photographic plate is particularly sensitive. In small telescopes the central part of this nebula appears dark but with a powerful telescope a faint light may be seen even in the central portion of the nebula. This is one of the most interesting and beautiful telescopic objects in the heavens.

It is in the general direction of the constellation of Lyra that our solar system is speeding at the rate of more than a million miles a day. This point toward which we are moving at such tremendous speed lies a little to the southwest of Vega, on the border between the constellations of Lyra and Hercules, and is spoken of as The Apex of the Sun's Way.

## August—Sagittarius

In the southern sky we have this month the constellation of Sagittarius, The Archer, which is just to the east of Scorpio and a considerable distance south of Lyra. It can be recognized by its peculiar form, which is that of a short-handled milk dipper, with the bowl turned toward the south and a trail of bright stars running from the end of the handle toward the southwest. This is one of the zodiacal groups which contain no first-magnitude stars, but a number of the second and third magnitude. It is crossed by the Milky Way, which is very wonderful in its structure at this point. Some astronomers believe that here—among the star-clouds and mists of nebulous light which are intermingled with dark lanes and holes, in reality dark nebulæ—lies the center of the vast system of stars and nebulæ in which our entire solar system is but the merest speck. Some of the grandest views through the telescope are also to be obtained in this beautiful constellation of Sagittarius, which is so far south that it is seen to better advantage in the tropics than in the mid-latitudes of the northern hemispheres.

# CHAPTER X. SEPTEMBER.

One of the most beautiful constellations of the northern hemisphere is Cygnus, The Swan, which is in the zenith in mid-latitudes about nine o'clock in the evening the middle of September. It lies directly in the path of the Milky Way which stretches diagonally across the heavens from the northeast to the southwest at this time. In Cygnus, the Milky Way divides into two branches, one passing through Ophiuchus and Serpens to Scorpio, and the other through Sagitta and Aquila to Sagittarius, to meet again in the southern constellation of Ara, just south of Scorpio and Sagittarius. On clear, dark evenings, when there is no moonlight, this long, dark rift in the Milky Way can be seen very clearly. In Cygnus, as in Ophiuchus, Scorpio, and Sagittarius we find wonderful star-clouds, consisting of numberless stars so distant from us and, therefore, so faint that they do not appear as distinct points of light except in the greatest telescopes. It is the combined light from these numberless stars that cannot be seen separately that produces the impression of stars massed in clouds of nebulous light and gives to this girdle of the heavens its name of the Milky Way. In Cygnus, as in a number of other constellations of both hemispheres, the Milky Way is crossed by dark rifts and bars and is very complicated in its structure. It is in Cygnus, also, that one may see with the aid of powerful telescopes the vast, irregular, luminous nebulæ, that are like great clouds of fiery mist. These nebulæ are of enormous extent, for they cover space that could be occupied by hundreds of stars.

September—Cygnus

Cygnus is a constellation that is filled with the wonders and mysteries of space and that abounds in beautiful objects of varied kinds. It is a region one never tires of exploring with the telescope. The principal stars in Cygnus form the well-known Northern Cross, with the beautiful, white, first-magnitude star Deneb, or Arided, as it is sometimes called, at the top of the cross, and Albireo, the orange-and-blue double star at the foot. Albireo, among all the pairs of contrasting hues, has the distinction of being considered the finest double star in the heavens for small telescopes. This star marks the head of The Swan, as well as the foot of the Northern Cross, and Deneb marks the tail of The Swan. A short distance to the southeast of Deneb, on the right wing of The Swan, is a famous little star, 61 Cygni, barely visible to the naked eye and forming a little triangle with two brighter stars to the east. This star has the distinction of being the first one to have its distance from the solar system determined. The famous mathematician and astronomer Bessel accomplished this difficult feat in the year 1838. Since that day, the distances of many stars have been found by various methods, but of all these stars only four or five are known to be nearer to us than 61 Cygni. Its distance is about eight light-years, so its light takes about eight years to travel the distance that separates it from the solar system. As a result, we see it not as it is tonight, but as it was at the time when the light now entering our eyes first started on its journey eight years ago. 61 Cygni is also a double star, and the combined light of the two stars gives forth only one-tenth as much light as our own sun. Most of the brilliant first-magnitude stars give forth many times as much light as the sun; but among the fainter stars, we find some that appear faint because they are very distant, and some that are faint because they are dwarf stars and have little light to give forth. To the class of nearby, feebly-shining dwarf stars 61 Cygni belongs. Deneb, on the other hand, is one of the giant stars, and is at an immeasurably great distance from the solar system.

Just south of Cygnus in the eastern branch of the Milky Way lie Sagitta, The Arrow, and Aquila, The Eagle. Not far to the northeast of Aquila is the odd little constellation of Delphinus, The Dolphin, popularly referred to as Job's Coffin. There will be no difficulty in finding this small star-group, owing to its peculiar diamond-shaped configuration. Its five principal stars are of the fourth magnitude. It is in the constellation of Delphinus that the most distant known object in the heavens is located. This is the globular star cluster known only by its catalogue number of N.G.C. 7006. It is estimated to be at a distance of 220,000 light-years from the earth.

Sagitta, The Arrow, lies midway between Albireo and the brilliant Altair in Aquila. The point of the arrow is indicated by the star that is farthest east; and the feather, by the two faint stars to the west. Like Delphinus, this constellation is very small and contains no objects of particular interest.

Altair (Flying Eagle) is the brilliant white star of the first magnitude in Aquila and is attended by two fainter stars, one on either side, at nearly equal distances from it. These two stars serve readily to distinguish this star, all three stars being nearly in a straight line. Altair is one of the nearer stars, its distance from the earth being about sixteen light-years. It gives forth about ten times as much light as the sun.

## September—Delphinus, Aquila and Sagitta

We cannot leave the constellation of Aquila without referring to the wonderful temporary star or nova, known as Nova Aquilæ No. 3 (because it was the third nova to appear in this constellation), which appeared in the position indicated on the chart upon the eighth of June, 1918. A few days previous to this date, there was in this position an extremely faint star, invisible to the naked eye and in small telescopes. This fact became known from later examinations of old photographs of this region that had been taken at the Harvard College Observatory, where the photographing of the heavens is carried on regularly for the purpose of having a record of celestial changes and happenings. Clouds prevented the obtaining of any photographs of this part of the heavens on the four nights preceding the eighth of June, but on this evening there shone in the place of the faint telescopic star, a wonderful temporary star, or nova, which was destined on the next evening to outshine all stars in the heavens, with the exception of the brightest of all, Sirius, which it closely rivaled in brilliancy at the height of its outburst. Within less than a week's time, this faint star in the Milky Way for some mysterious reason increased in brightness many thousandfold. Such outbursts have been recorded before, but on rare occasions, however. No star since the appearance of the nova known as Kepler's Star, in the year 1604, which at its greatest brilliancy rivaled Jupiter, shone with such splendor or attracted so much attention as Nova Aquilæ. In the year 1901, there appeared in the constellation of Perseus a star known as Nova Persei which at its brightest surpassed Vega, but its splendor was not as great as that of the wonderful nova of 1918.

It speaks well for the zeal and interest of amateur astronomers, as well as for their acquaintance with the stars, that Nova Persei was discovered by an amateur astronomer, Dr. Anderson, and that among the deluge of telephone calls and telegrams received at the Harvard College Observatory on the night of June 8th, announcing independent discoveries of the "new star," were many from non-professional astronomers.

Like all stars of this class, Nova Aquilæ No. 3 sank rapidly into oblivion. In a few weeks it was only a third-magnitude star; a few weeks more and it was invisible without a telescope. Many wonderful and interesting changes have been recorded in the appearance of this star, however, even after it became visible only in the telescope. Soon after its outburst it appeared to develop a nebulous envelope, as have other novas before it. It showed in addition many of the peculiarities of the nebulæ, though the central star remained visible as before the outburst.

Astronomers are still in doubt as to the cause of these outbursts, which certainly indicate celestial catastrophies of some form on a gigantic scale. All novas possess one characteristic in common—that of appearing exclusively in the Milky Way; and another characteristic is the development of a nebular envelope after the outburst of greatest brightness. In some cases temporary stars have been known to be variable in brightness for years before the great outburst. Such a star was Nova Aquila, for the examination of photographs of this region taken some years previous showed variations in its brightness for a period of thirty years at least.

Up to the beginning of this century only about thirty novas had been discovered. Since that date, thanks to the vigilance of the astronomers of today and to the aid of photography, more have been discovered than in all the preceding centuries. These outbursts of new stars appear to be not so rare as the earlier astronomers believed, though great outbursts as brilliant as that of Nova Aquila are very uncommon.

# CHAPTER XI. OCTOBER.

The constellations that will be found nearest the meridian in early October evenings are the circumpolar constellations Cepheus and Cassiopeia, and in the southern sky, Capricornus and Aquarius.

Cepheus, The King, and Cassiopeia, his Queen, of whom we shall have more to say later in connection with the constellations of Andromeda and Perseus, sit facing the north pole of the heavens opposite Ursa Major, The Great Bear, familiar to us under the name of The Big Dipper. The foot of Cepheus rests upon the tail of the Little Bear, and the star farthest north in the diagram is in the left knee. The head is marked by a small triangle of faint stars, shown in the diagram. One of these three faint stars—the one farthest east—known as Delta Cephei, is a very remarkable variable star, changing periodically in brightness every five and one-third days. Its name has been given to a large class of variable stars—the Cepheid variables—that resemble Delta Cephei in being giant suns, faint only because they are at very great distances from the earth, and varying in brightness with the greatest regularity in periods that range from a few hours to several weeks. It has been found that the longer the period of light change the greater is the star in size and brightness. The Cepheids of longest period are 10,000 times more brilliant than our own sun. Cepheus contains no very bright or conspicuous stars. Alpha Cephei, the brightest star in the group, marks The King's right shoulder. It is the star farthest to the west in the diagram, and is only a third-magnitude star.

October—Cassiopeia and Cepheus

Cassiopeia is a constellation with which every one in the northern hemisphere should be familiar, owing to its very distinctive W-shape and its far northern position, which brings it conspicuously into view throughout the clear fall and winter evenings. Cassiopeia is pictured in all star atlases that show the mythological figures, with her face toward the north pole. The stars in the W outline the body. Alpha, the star farthest south in the diagram marks the breast of Cassiopeia. Her head and uplifted hands are represented by faint stars south of Alpha. This star is occasionally referred to by its Arabic name of Schedir. Beta, the leader of all the stars in the W in their daily westward motion, is also known by an Arabic name, Caph.

In the constellation of Cassiopeia there appeared in the year 1572 A.D., a wonderful temporary star which suddenly, within a few days' time, became as brilliant as the planet Venus and was clearly visible in broad daylight. This star is often referred to as Tycho's Star, because it was observed, and its position very accurately determined, by Tycho Brahe, one of the most famous of the old astronomers. This star remained visible to every one for about sixteen months, but it finally faded completely from view, and it is believed that a faint, nebulous red star, visible only in the telescope and close to the position recorded by Tycho, represents the smoldering embers of the star that once struck terror to the hearts of the superstitious and ignorant among all the nations of Europe, who took it to be a sign that the end of the world was at hand.

Both Cassiopeia and Cepheus lie in the path of the Milky Way, which reaches its farthest northern point in Cassiopeia and passes from Cepheus in a southerly direction into the constellation of Cygnus.

## October—Aquarius and Capricornus

Turning now to southern skies, we find on and to the west of the meridian at this time the rather inconspicuous zodiacal constellation of Capricornus, The Goat. It contains no stars of great brightness and is chiefly remarkable for the fact that it contains one of the few double stars that can be seen without the aid of a telescope. The least distance in the heavens that the unaided human eye can separate is about four minutes of arc. The star Alpha in Capricornus is made up of two stars separated by a distance of six minutes of arc, so that any one can readily see that it consists of two stars very close together. This star, Alpha, will be found in the extreme western part of the constellation, and can best be located in conjunction with the star Beta, which is slightly brighter and lies but a short distance almost due south of Alpha, the two stars standing somewhat alone in this part of the heavens.

To the north and east of Capricornus we find Aquarius, which is also a zodiacal constellation. Aquarius is the Water-Bearer, and the water jar which he carries is represented by a small, but distinct, Y of stars from which flows a stream of faint stars toward the southeast and south. Aquarius, like Capricornus, is a rather uninteresting constellation, as it is made up of inconspicuous third- and fourth-magnitude stars. The entire region covered by these two groups of stars is remarkably barren, since it contains not a single first- or even second-magnitude star and little to attract the observer's eye.

To relieve the barrenness of this region, there appears just to the south of Aquarius and southeast of Capricornus, sparkling low in the southern sky on crisp October evenings, the beautiful first-magnitude star Fomalhaut in the small southern constellation of Piscis Australis, The Southern Fish. This star is the farthest south of all the brilliant first-magnitude stars that can be seen from the middle latitudes of the Northern Hemisphere. The constellation in which it lies is so close to the southern horizon in our latitudes that it cannot be seen to any advantage, and it is at best very inconspicuous, containing no other objects of interest. Fomalhaut cannot be mistaken for any other star visible at this time of year in the evening, since it stands in such a solitary position far to the south. At the time of which we are writing it will be found a few degrees east of the meridian.

# CHAPTER XII. NOVEMBER.

Directly south of Cassiopeia and Cepheus, the circumpolar constellations with which we became acquainted last month, and almost overhead in our latitudes in the early evening hours of November, lie Pegasus, The Winged Horse, and Andromeda, The Woman Chained.

According to the legend, Cepheus was king of Ethiopia, and Cassiopeia was the beautiful, but vain, queen who dared to compare herself in beauty with the sea-nymphs. This so enraged the nymphs that, as a punishment for her presumption, they decided to send a terrible sea-monster to ravage the coast of the kingdom. The king and queen, upon consulting the oracle, found that the only way to avert this calamity would be to chain their daughter Andromeda to the rocks and permit the monster to devour her.

November—Andromeda and Pegasus

As the story goes, the valiant hero, Perseus, chanced to be riding through the air on his winged horse and saw, far beneath him the beautiful maiden chained to the rocks and the frightful monster approaching to devour her. He immediately went to the rescue, and, after a terrible struggle with the monster, succeeded in overpowering him and thus saved the maiden from a dreadful fate. Perseus and the fair Andromeda were married shortly afterward, and at the end of a happy life the pair were transferred to the heavens. Cassiopeia, the vain queen, was ordered to be bound to a chair and, with the king Cepheus at her side, to be swung continually around the north pole of the heavens that she might be taught a lesson in humility.

The constellation Cetus, representing the sea-monster, will be found to the southeast and south of Pisces, The Fishes, which lie south of Andromeda and Pegasus.

The Great Square in Pegasus is the most conspicuous configuration of stars to be seen in the heavens in autumn evenings. The star that marks the northeastern corner of The Great Square belongs to the constellation of Andromeda and marks the head of the maiden, who is resting upon the shoulders of Pegasus, The Winged Horse. The three bright stars nearly in a straight line outline the maiden's body, Alpha, or Alpheratz, as it is called, being the star in the head, Beta or Mirach in the waist, and Gamma or Almach in the left foot. The last-named star, which is farthest to the northeast in the diagram, was, in the opinion of the noted astronomer

Herschel, the finest double star in the heavens. The two stars into which the telescope splits it are of the beautifully contrasted shades of orange and sea green.

A second most interesting object in Andromeda and one of the finest in the entire heavens is The Great Andromeda Nebula, which is faintly visible without the aid of a telescope as a hazy patch of light. It is believed that in reality this nebula is a great universe composed of many thousands of stars so distant that no telescope can show the individual members and that the light from it takes many thousands of years to span the abyss that separates it from the solar system. Some magnificent photographs of The Great Andromeda Nebula have been taken with powerful telescopes. It is through the use of photography that the nebulæ can best be studied, for a photographic plate after long exposure, reveals a wonderful detail in the structure of these objects that the human eye fails to see. On a clear, dark evening one may find The Great Andromeda Nebula by the aid of two faint stars with which it makes a small triangle, as shown in the chart. This nebula is the only one of the spiral nebula that can be seen in these latitudes without the aid of a telescope, though there are several spiral nebulæ in the southern heavens that can be thus seen.

Lying to the northwest of The Great Square in Pegasus are a number of faint stars that outline the shoulders and head of the winged steed, while the stars to the southwest of the square outline his forelegs. The creature is represented without hind quarters in all star atlases. The space within The Great Square contains no bright stars, and as a result, the outline of the square stands out with great distinctness. There are, in fact, no stars of the first magnitude in either Pegasus or Andromeda, though there are a number of the second and third magnitude which very clearly show the distinctive forms of these two star-groups.

Pisces, The Fishes, the constellation just south of Andromeda and Pegasus, is the first of the twelve zodiacal constellations. It consists of the southern fish, lying in an east-to-west direction, and the northern fish, lying nearly north and south, the two touching at the southeastern extremity of the constellation.

November—Pisces

There is in Pisces not a single bright star, and its only point of interest is to be found in the fact that it contains the point, marked by the cross and letter V in the diagram, that is known variously as "the vernal equinox," "the equinoctial point" and "The First Point in Aries." This is a very important point of reference in the heavens, just as the meridian of Greenwich is for the earth, and it marks the point where the sun crosses the equator going north in the spring. Owing to the Precession of the Equinoxes, as it is called, this point is gradually shifting its position westward through the zodiacal constellations at a rate that will carry it completely around the heavens through the twelve zodiacal groups in a period of 25,800 years. Since the beginning of the Christian era, this point has backed from the constellation of Aries, which lies just east of Pisces, into Pisces, though it still retains its name of "The First Point in Aries."

## CHAPTER XIII. DECEMBER.

The eastern half of the sky on early December evenings is adorned with some of the finest star-groups in the heavens; but as we are considering for each month only the constellations that lie on or near the meridian in the early evening hours, we must turn our eyes for the present from the sparkling brilliants in the east to the stars in the less conspicuous groups of Aries, The Ram, and Cetus, The Whale. We will also become acquainted this month with the beautiful and interesting constellation of Perseus, the hero of mythical fame to whom we referred last month in connection with the legend of Cepheus and Cassiopeia. Cetus, you will recall, represents the sea-monster sent to devour Andromeda, the daughter of Cepheus and Cassiopeia. We have included the constellation of Andromeda in our diagram for this month, since it is so closely associated in legend with the constellations of Perseus and Cetus, though we also showed it last month.

The brightest star in Perseus, known as Alpha Persei, is at the center of a curved line of stars that is concave or hollow toward the northeast. This line of stars is called the Segment of Perseus, and it lies along the path of the Milky Way, which passes from this point in a northwesterly direction into Cassiopeia. According to the legend, Perseus, in his great haste to rescue the maiden from Cetus, the monster, stirred up a great dust, which is represented by the numberless faint stars in the Milky Way at this point. The star Alpha is in the midst of one of the finest regions of the heavens for the possessor of a good field-glass or small telescope.

A short distance to the southwest of Alpha is one of the most interesting objects in the heavens. To the ancients, it represented the baleful, winking demon-eye in the head of the snaky-locked Gorgon, Medusa, whom Perseus vanquished in one of his earliest exploits and whose head he carried in his hand at the time of the rescue of Andromeda. To the astronomers, however, Algol is known as Beta Persei, a star that has been found to consist of two stars revolving about each other and separated by a distance not much greater than their own diameters. One of the stars is so faint that we speak of it as a dark star, though it does emit a faint light. Once in every revolution the faint star passes directly between us and the bright star and partly eclipses it, shutting off five-sixths of its light. This happens with great regularity once in a little less than three days. It is for this reason that Algol varies in brightness in this period. There are a number of stars that vary in brightness in a similar manner. Their periods of light-change are all very short, and the astronomers call them eclipsing variables. At its brightest, Algol is slightly brighter than the star nearest to it in Andromeda, which is an excellent star with which to compare it.

December—Perseus, Aries and Cetus

Perseus is another one of the constellations lying in the Milky Way in which temporary stars or novas have suddenly flashed forth. At the point indicated by a cross in the diagram, Dr. Anderson, an amateur astronomer of Scotland, found on February 21, 1901, a new star as brilliant as the pole-star. On the following day it became brighter than a star of the first magnitude. A day later it had lost a third of its light, and in a few weeks it was invisible without the aid of a telescope. In a year it was invisible in all except the most powerful telescopes. With such telescopes, it may still be seen as a very faint nebulous light.

Triangulum and Aries are two rather inconspicuous constellations that lie on, or close to, the meridian at this time. There is nothing remarkable about either of these groups, except that Aries is one of the twelve zodiacal constellations. Some centuries ago, the sun was to be found in Aries at the beginning of spring and the position it occupies in the sky at that time was called the First Point in Aries. As this point is slowly shifting westward, as we have explained elsewhere, the sun is now to be found in Pisces, instead of Aries at the beginning of spring and does not enter Aries until a month later. Pisces was one of the constellations for November and we showed in that constellation the present position of the sun at the beginning of spring.

Two stars in Aries—Alpha and Beta—are fairly bright, Alpha being fully as bright as the brightest star in Andromeda. Beta lies a short distance to the southwest of Alpha, and a little to the southwest of Beta is Gamma, the three stars forming a short curved line of stars that distinguishes this constellation from other groups. The remaining stars in Aries are all faint.

Just south of Aries lies the head of Cetus, The Whale. This is an enormous constellation that extends far to the southwest, below a part of Pisces, which runs in between Andromeda and Cetus. Its brightest star, Beta, Diphda, or Deneb Kaitos, as it is severally called, stands quite alone not far above the southwestern horizon. It is almost due south of Alpha Andromedæ, the star in Andromeda farthest to the west, which it exactly equals in brightness. The head of Cetus is marked by a five-sided figure composed of stars that are all faint with the single exception of Alpha, which is fairly bright, though inferior to Beta or Diphda.

Cetus, though made up chiefly of faint stars, and on the whole uninteresting, contains one of the most peculiar objects in the heavens, the star known as Omicron Ceti or Mira (The Wonderful). This star suddenly rises from invisibility nearly to the brightness of a first-magnitude star for a short period once every eleven months. Mira was the first known variable star. Its remarkable periodic change in brightness was discovered by Fabricius in the year 1596, so its peculiar behavior has been under observation for three hundred and twenty-five years. It is called a long-period variable star, because its variations of light take place in a period of months instead of a few hours or days, as is the case with stars such as Algol. Mira is not only a wonderful star, it is a mysterious star as well, for the cause of its light-changes are not known, as in the case of Algol where the loss of light is produced by a dark star passing in front of a brighter star. Mira is a deep-red star, as are all long-period variable stars that change irregularly in brightness. It is visible without a telescope for only one month or six weeks out of the eleven months. During the remainder of this time, it sometimes loses so much of its light that it cannot be found with telescopes of considerable size. Its periods of light-change are quite variable as is also the amount of light it gains at different appearances.

It is believed that the cause of the light-changes of Mira is to be found within the star itself. It has been thought that dense clouds of vapors may surround these comparatively cool, red stars and that the imprisoned heat finally bursts through these vapors and we see for a short time the glowing gases below; then the vapors once more collect for a long period, to be followed by another sudden outburst of heat and light.

It is interesting to remember in this connection that our own sun has been found to be slightly variable in the amount of light and heat that it gives forth at different times, and the cause of its changes in light and heat are believed to lie within the sun itself.

## CHAPTER XIV. STARS OF THE SOUTHERN HEMISPHERE.

As one travels southward from the mid-latitudes of the northern hemisphere into the tropics our familiar circumpolar constellations sink lower and lower in the northern heavens and strange and unfamiliar star-groups rise gradually above the southern horizon. If we make our southward journey in the winter months the first of the southern constellations to come fully into view is the small star-group just south of Lepus known as Columba (The Dove), whose brightest star Phact is of the second magnitude. A line drawn from Procyon to Sirius and extended as far again brings us to this star and a line from Betelgeuze to Sirius extended an equal distance brings us to Zeta Argus which is equal to Phact in brightness. The two lines intersecting at Sirius make the "Egyptian X" as it is called.

Magnificent, blue-white Canopus, the most brilliant star in the heavens next to Sirius, a veritable diamond sparkling low in the southern sky, now commands our unqualified admiration. Canopus lies about 35° south of Sirius and is invisible north of the 37th parallel of latitude. At nine o'clock in the evening of February 6th it can be seen just above the southern horizon in that latitude and is then a conspicuous object in Georgia, Florida and the Gulf States.
writes Moore of Canopus in "Lalla Rookh."

Along the Nile Canopus was an object of worship as the god of waters. At the time of their erection, 6400 B.C., a number of temples in Upper Egypt were oriented so as to show Canopus at sunrise at the autumnal equinox, and other temples erected many centuries later were oriented in a similar manner. In China, as late as 100 B.C., and in India also Canopus was an object of worship.

The astronomer tells us that Canopus is immeasurably distant from the earth. It has been estimated to be forty thousand times more luminous than our sun.

Canopus is located in the constellation of Argo Navis which is the largest and most conspicuous constellation in the heavens. In addition to Canopus it contains a number of second- and third-magnitude stars and is subdivided for convenience into Puppis, The Prow; Carina, The Keel; and Vela, The Sails. Huge as it is, Argo Navis represents only half of a ship for the stern is lacking. According to the legend this ship was built by Argos, aided by Pallas Athene, for Jason, the leader of the expedition of the fifty Argonauts who sailed from Greece to Colchis in search of the golden fleece. Pallas Athene placed in the bow of the ship a piece of timber from the speaking oak of Dodona to guide the crew and warn them of dangers and after the voyage the ship was supposed to have been placed in the heavens.

Southern Constellations—1. In February

In Argo Navis is one of the finest telescopic objects in the heavens, the Keyhole Nebula, as it is usually called, from a peculiar-shaped dark patch in its brightest part. On the border of this nebula is the deep-red Wonder Star of the southern hemisphere, Eta Argus, which varies suddenly and unexpectedly in brightness between the seventh and first magnitudes. In 1843 it burst forth with a splendor rivaling Sirius and maintained this brilliancy for nearly ten years and then slowly waned in brilliancy until it disappeared to the unaided eye in 1886. The surrounding nebula also seems to share in its peculiar fluctuations of brightness. Eta Argus is now a star of the seventh magnitude and since it is still varying fitfully in brightness it is believed that the history of its light-changes is not complete.

Among the constellations of the southern heavens near the meridian in February we see in addition to Argo Navis the constellations of Dorado, The Goldfish; Hydrus, The Serpent, and Tucana, The Toucan. Though insignificant in appearance Dorado contains what was described by Sir John Herschel as one of the most extraordinary objects in the heavens, a worthy rival of The Great Orion Nebula and in some respects very similar to it, The Great Looped Nebula, "the center of a great spiral." In Dorado also is located The Greater Magellanic Cloud which looks like a detached portion of the Milky Way though it is far removed from it. To the naked eye it resembles a small white cloud about 4° in extent. In the telescope it bears a close resemblance to a typical portion of the Milky Way. A similar formation known as The Lesser Magellanic Cloud is located in Hydrus. It has been estimated that the distance of The Lesser Cloud is 80,000 light-years and that it is receding from us.

In Tucana is located one of the finest globular star clusters in the heavens, known as 47 Tucanæ. This cluster and Omega Centauri, a globular star cluster in Centaurus, are the two nearest of all the globular clusters. They are distant from the earth about 22,000 light-years and it is known that the combined light of the thousands of stars of which each cluster is composed is about one million times that of our own sun.

In the western sky in the southern hemisphere in February may be seen the brilliant, white, first-magnitude star Achernar in the river Eridanus, the long, winding constellation that, we recall, starts near the brilliant Rigel in Orion and disappears from the view of northern observers below the southern horizon, extending its course far into the southern hemisphere. Achernar means "The End of the River" and this is nearly its position in the constellation.

Though Argo Navis is the largest and most important constellation of the southern hemisphere, Crux, The Southern Cross, far-famed in story and legend as well as for its historical associations, is beyond a doubt the most popular.

The best time to view the Southern Cross is in June or July when it is near the meridian. It is not seen to advantage in the months of January or February. It then lies on its side and close to the horizon and therefore is dimmed by atmospheric haze so that it almost invariably is a disappointing object to the tourist from the north who usually views it for the first time in one of these months. The Cross is viewed to advantage in the latitude of Rio or Valparaiso and it is best seen from the Straits where it rides high overhead. It is not seen to advantage from the latitudes of Cuba or Jamaica. It is small, only 6° in extent from north to south and less in width and it lies in the most brilliant portion of the Milky Way which is here a narrow stream only three or four degrees wide. Directly below the Cross is the noted Coal Sack, apparently a yawning chasm in the midst of its brilliant surroundings though probably in reality a dark nebula. Viewed with the telescopes a number of stars are to be seen projected on this dark background.

The Southern Cross is to the inhabitants of the southern hemisphere what the Big Dipper is to those who dwell in the northern hemisphere—an infallible timepiece. The upright of the Cross points toward the south pole of the heavens which lies in a region where there is a singular dearth of bright stars, the nearest star to the south pole being a faint fifth-magnitude star called Sigma Octantis. When seen in the southeast or southwest the Cross lies on its side, but when passing the meridian it stands nearly upright. Humboldt, referring to this fact, says:

"How often have we heard our guides exclaim in the savannahs of Venezuela and in the desert extending from Lima to Truxillo, 'Midnight is past, the Cross begins to bend.'"

By the explorers of the sixteenth century the Cross was taken as a sign of heaven's approval of their attempt to establish the Christian religion in the wilds of the New World. This thought is beautifully expressed in Mrs. Hemans' lines in "The Cross of the South."
"But to thee, as thy lode-stars resplendently burnIn their clear depths of blue, with devotion I turnBright Cross of the South! and beholding thee shine,Scarce regret the loved land of the olive and vine.Thou recallest the ages when first o'er the mainMy fathers unfolded the ensign of Spain,And planted their faith in the regions that seeIts imperishing symbol ever blazoned in thee."

Alpha Crucis, the brightest star in Crux, is at the foot of the Cross. It consists in reality of two second-magnitude stars forming a beautiful double while a third fifth-magnitude star one and one-half minutes of arc distant makes with this pair a combination similar to our Mizar and Alcor of the Big Dipper though the separation is not great enough to be visible to the naked eye. The second-magnitude star at the head of the Cross is a deep orange in color and the two stars that mark the ends of the cross-arm are white third-magnitude stars.

One of the finest constellations of the southern hemisphere is Centaurus, The Centaur, which surrounds Crux on the north and is more than 60° in length. Its center lies about 50° south of Spica in Virgo and below the tail of Hydra. Alpha Centauri, its brightest star and the nearest star to the solar system, four and one-third light-years away, is a golden-yellow double star that forms with the star Beta Centauri on the west a configuration similar to that of Castor and Pollux in Gemini, only one that is far more striking because of the superior brilliancy of the stars. Alpha Centauri lies in the Milky Way and transits the meridian at the same time with Arcturus though it cannot be seen north of the 29th parallel. Alpha Centauri, like Canopus, was an object of worship in Egypt and a number of temples in northern Egypt were oriented to its emergence from the sun's rays in the morning at the autumnal equinox, between 3000 and 2575 B.C.

North of Centaurus is the constellation Lupus, The Wolf, which is also crossed by the Milky Way. According to one myth Lupus is held in the right hand of the Centaur as an offering upon the altar which is represented by the constellation of Ara next to Centaurus on the east. Ara also is crossed by the Milky Way. Neither Lupus nor Ara contain any objects that are worthy of special attention.

Triangulum Australe, The Southern Triangle, a little to the southeast of Alpha Centauri, is far more conspicuous than the Triangulum of the northern hemisphere.

The accompanying charts give two views of these principal southern constellations that lie within 40° of the south pole of the heavens and that are below the horizon in 40° north latitude. The first of these charts shows the constellations that are nearest the meridian in the early evening hours in February. Canopus in Argo Navis and the Greater Magellanic Cloud then lie close to the meridian. Argo Navis with its subdivisions Puppis, Vela and Carina are found east of the meridian lying directly in the path of the Milky Way, which stretches diagonally across the sky from the northwest to the southeast. Far over in the southeast appears Crux, the Southern Cross, also directly in the path of the Milky Way. In the western heavens may be seen the Lesser Magellanic Cloud in Hydrus, brilliant Achernar in Eridanus and the inconspicuous star-group of Tucana.

In the early evening hours of July we find as shown on the second chart, Alpha and Beta Centauri in Centaurus close to the meridian, Lupus due north of Centaurus, Ara and Triangulum Australe in the southeast and Crux in fine position for observation just west of the meridian. Carina of Argo Navis lies to the southwest of Crux. The Milky Way now arches magnificently across the heavens from Carina through Crux, Centaurus and Lupus and Ara to the zodiacal groups of Scorpio and Sagittarius in the northeast.

In the northern part of the heavens, as seen from the southern hemisphere, appear the familiar zodiacal constellations that we of the northern hemisphere find south of the zenith, as well as the constellations of Orion, Lepus and Canis Major, Hydra, Corvus and Crater, Ophiuchus and Serpens and Aquila, all finely in view in their appropriate seasons.

It is only our familiar circumpolar constellations of the north—The Two Bears, Draco, Cassiopeia, and Cepheus, Andromeda and Perseus and Auriga that are invisible in mid-latitudes of the southern hemisphere just as the constellations shown in the diagrams, and a few additional groups such as Pavo, Grus, Phoenix, Apus, Mensa and Volans which we have not shown, lie hidden from view beneath the southern horizon in mid-latitudes of the northern hemisphere.

The northern visitor to the southern hemisphere familiar with the constellations of his own land is filled with a queer sensation of being in topsy-turvydom as he sees familiar Orion standing on his head and all of the zodiacal constellations passing in their daily motions to the north instead of to the south of his zenith while by day the sun passes across the northern part of the heavens and culminates north instead of south of his zenith. He misses the familiar Dippers of his own land and searches in vain for a pole-star in the unfamiliar circumpolar regions of the south.

# CHAPTER XV. THE MILKY WAY OR GALAXY.

"Broad and ample road whose dust is gold,And pavement stars, as stars to thee appearSeen in the galaxy, that milky wayWhich nightly as a circling zone thou seestPowder'd with stars."

—MILTON, Paradise Lost.

On clear, winter evenings one may see a portion of the zone of the Milky Way, which encircles the heavens, arching magnificently across the heavens as it passes from Cassiopeia and Cepheus in the northwest, through Perseus and Auriga and the eastern part of Taurus, across the feet of Gemini, between Canis Minor and Orion and through the eastern part of Canis Major to the southern horizon.

At this point it passes beyond our range of vision into the star-groups of Puppis, Vela and Carina, subdivisions of the huge southern constellation of Argo Navis. It reaches its greatest distance south of the celestial equator and also attains its greatest brilliancy in Crux, the far-famed Southern Cross. From here it turns northward once more, passing into Centaurus, Musca, Circinus Ara and Lupus constellations of the southern hemisphere and comes within our range of vision again in Sagittarius and Scorpio. Here the Milky Way divides into two branches, though some astronomers now believe that this apparent division into two branches is due to the presence of an enormous cloud of non-luminous matter lying along the course of the Milky Way at this point, similar in its nature to the dark "holes" and "caves" and streaks that appear in all portions of the Milky Way and most noticeably athwart its course in Argo and Centaurus.

One of these branches of the Milky Way passes from Sagittarius through Aquila to Cygnus and the other through Scorpio, Ophiuchus and Serpens to Cygnus, the two extending diagonally across the heavens in the late summer and early fall evenings from the northeast to the southwest. From Cygnus, the Milky Way passes into Cepheus and Cassiopeia and thus completes its circuit of the heavens.

It is not seen to advantage in spring or early summer evenings because it then rests nearly on the horizon. Its plane is inclined about 63° to the celestial equator and its poles lie in the constellations of Coma Berenicis and Cetus. These are the two points that lie farthest from the Milky Way.

The Milky Way has been called the groundwork of the universe. By far the greater number of all the stars are crowded towards its plane in the form of an enormous flattened disk or lens.

Our solar system, it has been estimated, lies close to the plane of the Milky Way and at a distance of some 50,000 or 60,000 light-years from its center. The diameter of the Milky Way as deduced from Dr. Harlow Shapley's work on globular star clusters is about 300,000 light-years in extent, or ten times greater than the limit set some years ago.

The apparent crowding together of the stars into dense clouds in the Milky Way is partly an effect due to our position in the Milky Way. When we look at the heavens in a direction at right angles to this plane we find comparatively few stars lying along our line of vision because the stars are actually fewer in number in this direction. If we look along the plane of the Milky Way, however, we see to a greater distance through an enormous depth of stars. Though individual stars may not be much closer together in the Milky Way than they are outside of it, there are on the whole more of them and the effect of greater density is produced.

Father Hagen of the Vatican Observatory, who has for years made a study of the dark clouds of obscuring matter and dark nebulæ that abound in space, has found evidence of the existence of many vast clouds of dark obscuring matter over the entire heavens above and below the plane of the Milky Way as well as surrounding the Milky Way in its own plane. The existence of such clouds of non-luminous matter would account partly for the comparative fewness of stars in space outside of the plane of the Milky Way since many stars would be concealed from our eyes by these obscuring clouds. There is, however, in addition, an actual crowding of all the visible stars toward this plane.

The peoples of all ages have honored the Milky Way in story and legend. It has been universally referred to as The Sky River and The Pathway of Souls. To the little Hiawatha, we remember, the "wrinkled old Nakomis"
"Showed the broad white road in heavenPathway of the ghosts, the shadows,Running straight across the heavens,Crowded with the ghosts, the shadows.To the Kingdom of PonemahTo the land of the hereafter."

In The Galaxy, Longfellow thus describes the Milky Way:
"Torrent of light and river of the airAlong whose bed the glimmering stars are seenLike gold and silver sands in some ravineWhere mountain streams have left their channels bare!"

In Sweden, where the Milky Way arches high through the zenith in winter, it is called the Winter Street, and Miss Edith Thomas writes thus beautifully of it in her poem entitled, "The Winter Street":

Faintly it gleams, like a summer roadWhen the light in the west is sinking low,Silent with star dust! By whose abodeDoes the Winter Street in its windings go?
And who are they, all unheard and unseen—O who are they, whose blessèd feetPass over that highway smooth and sheen?What pilgrims travel the Winter Street?
Are they not those whom here we missIn the ways and the days that are vacant below?As the dust of that Street their footfalls kissDoes it not brighter and brighter grow?"

Beautiful indeed are these poetic fancies but none of them picture even remotely the awe-inspiring grandeur of the Milky Way as it actually exists.

Millions upon millions of far distant suns equal to or surpassing our own sun in brilliancy are gathered within this vast encircling zone of the heavens, their combined light giving to the naked eye the impression of a milky band of light. Nine-tenths of all the stars, it has been estimated, lie close to the plane of the Galaxy, as well as all the vast expanses of luminous gaseous nebulæ and clouds of dark obscuring matter all seemingly intermingled in chaotic confusion; yet law and order govern the motions of all. Here also are the great moving star clusters such as the Pleiades and the Hyades and all of the brilliant "Orion" stars.

The structure of the Milky Way is not clearly understood but many astronomers believe there is evidence that it takes the form of a vast spiral nebula along whose arms the stars pass to and fro.

Beyond the Milky Way at enormous distances of many thousands of light-years, but apparently influenced by it, lie the globular star-clusters and the spiral nebulæ. The spirals appear to avoid the plane of the Milky Way for they are receding in the direction of its poles at high velocities; the globular clusters on the other hand are drawing in toward the Milky Way on either side, and in time will cross it.

Whether these objects external to the Milky Way form with it one enormous universe or whether the spiral nebulæ are in turn galaxies or "island universes," as the astronomer calls them, similar in form and structure to our own galaxy and at inconceivably great distances of millions of light-years from it, is still one of the riddles of the universe which the astronomers are attempting to solve.

# CHAPTER XVI. THE SURFACE OF THE SUN.

The visible surface of the sun is called the photosphere. Even the smallest telescopes will show its peculiar "rice-grain" structure, consisting of intensely brilliant flecks or nodules about 500 miles in diameter, which can be resolved by the more powerful telescopes into smaller particles about 100 miles in diameter, against a darker background. It has been estimated that these bright nodules or rice-grains occupy only one-fifth of the total surface of the sun, yet radiate three-fourths of the total light.

It is generally believed that the "rice grains" are the summits of highly heated columns of gas, arising from the sun's interior, and that the darker portions between are cooler descending currents.

It is well known that the photosphere or visible surface of the sun appears to be much brighter in the center of the disk than near its circumference. This is due entirely to its gaseous nature and to the fact that it is surrounded by an atmosphere of dense enveloping cooler gases. Rays from the center of the disk travel a shorter distance through this atmosphere than the rays from the rim and therefore are absorbed less by surrounding gases. We look further down into the sun's interior near the center of the disk than in the direction of its circumference and so the light appears more intense there.

The photosphere is the region where sun-spots appear and they are found in zones extending from 8° to 35° on either side of the solar equator, never appearing exactly at the equator or near the poles.

The disturbances that produce sun-spots and many allied phenomena occur cyclically in periods of eleven years on the average. The first outburst of the disturbance is manifested by the appearance of sun-spots in high solar latitudes. These break out and disappear and break out again with increased vigor, working gradually downward toward the solar equator, the maximum spottedness for a given period occurring in solar latitude about 16°. The disturbance finally dies out within 8° or 10° of the equator, but even before one cycle of disturbance has entirely passed away a new cycle has broken forth in high latitudes. So during the period of minimum spottedness there are four distinct belts, two in low latitudes, due to the dying disturbance, and two in high latitudes, due to the new disturbance. At sun-spot maximums there are two well-marked zones of great intensity, approximately 16° north and south of the sun's equator.

Sun-spots are solar cyclones, occurring usually in groups, though large single spots appear less frequently. Each spot is quite sharply divided into an umbra and a penumbra. The umbra is the darker central portion, the funnel of the whirling cyclone, and the penumbra is composed of the outspreading gases, and is less dark than the umbra. The peculiar "thatch-straw" structure of the penumbra is due, it is believed, to the fact that the columns of gases that usually rise vertically from the sun's interior and from the "rice grains" of the photosphere are drawn into a horizontal position by the whirling motion that exists in the penumbra regions of a sun-spot and therefore we get a longitudinal rather than a cross sectional view of them.

The umbra of a sun-spot is anywhere from a few hundred miles to fifty thousand miles in diameter, frequently exceeding the earth in size, while the penumbra occasionally reaches a diameter of two hundred thousand miles. Sun-spots of exceptional size can be seen even without the aid of a telescope.

The darkness of sun-spots is only by comparison with their more brilliant background. Owing to the rapid expansion and cooling of gases the temperature in sun-spot regions is far below the normal solar temperature of 6,000° Centigrade, lying between 3,000° and 4,000° Centigrade. At this temperature it is possible for the more refractory chemical compounds to form, the oxides and the hydrides, and the spectra of sun-spots reveal the presence of titanium oxide and magnesium and calcium hydride. At the higher solar temperatures that exist elsewhere in the photosphere and in its overlying gaseous envelopes all chemical elements occur in a free state, intermingling as incandescent vapors without the formation of any chemical compounds.

Strong magnetic fields exist in sun-spot regions and magnetic storms in our own atmosphere frequently accompany the appearance of exceptionally large sun-spots.

Directly above the photosphere of the sun lies the "reversing layer," which is about five hundred miles in depth and is composed of the incandescent vapors of all the chemical elements that exist on the sun, which are also the same familiar elements that exist on the earth, with the exception of coronium, the unknown element in the solar corona, there is no element in the sun that has not been found on our own planet.

The "reversing layer" receives its name from the fact that it reverses the solar spectrum. It produces by its absorption of the rays of light from gases below the dark absorption lines found in the spectrum that serve to identify all the elements existing in the sun. During the time immediately preceding and following a total eclipse of the sun this reversing layer produces what is known as the flash spectrum. When the photosphere, which gives the bright continuous background of the solar spectrum, is concealed by the moon, the normally dark lines of the reversing layer—dark only by contrast with the bright background—become momentarily intensely bright lines against a dark background. The flash spectrum only lasts a second or so, as the reversing layer itself is soon covered by the moon.

Just above the reversing layer lies the chromosphere, which is between five thousand and ten thousand miles in depth. Many of the gaseous vapors of the reversing layer are found in the chromosphere, thrown there continually by the vast upheavals of gases that are constantly disturbing the surface of the sun. The greater the solar activity the more is the chromosphere charged with the vapors of the lower strata of the sun's atmosphere. The gases that are most characteristic of the chromosphere, however, are the incandescent gases of hydrogen and calcium, which give it the pink or reddish tinge so noticeable during total solar eclipse. Helium is also found in great abundance in the solar chromosphere.

Shooting upward from the photosphere with the tremendous velocity of one hundred or more miles per second, can be seen at all times, by properly screening off the light from the photosphere, the vast solar eruptions known as the prominences. These are composed chiefly of hydrogen and calcium gas, though other elements also appear, especially near the bases of the prominences. Prominences are of two varieties, the quiescent, or cloud-like prominences, that float high above the solar surface for days at a time in some instances and resemble terrestrial clouds in form, and the eruptive, or metallic prominences, that dart up from the surface of the sun in an infinite variety of forms that may be entirely changed in the short interval of fifteen or twenty minutes.

These eruptive prominences usually attain heights of thirty or forty thousand miles on the average, but exceptional prominences reach heights of more than one hundred thousand miles and in a few rare cases have reached elevations of over five hundred thousand miles, or more than one-half of the solar diameter.

Prominences are the most spectacular and beautiful of all solar phenomena, with the possible exception of the solar corona, which is the outermost of all the solar envelopes and also the most tenuous. The extent of the corona is enormous. Its outer streamers extend usually to distances of several million miles from the center of the sun. Measurements of the coronal light during total eclipses of the sun have shown that its intensity is only about one-half that of full moonlight, and it seems almost impossible to devise methods for detecting it, except during total eclipses, on account of the extreme faintness of its light.

The sun, it is now known, is surrounded by a strong magnetic field in addition to the magnetic fields that exist in sun-spots. The cycle of sun-spot change is attended by marked changes in many forms of solar activity. The frequency of outbursts of eruptive prominences, the brightness and form of the corona, magnetic storms and weather changes on the earth are all closely associated with the sun-spot cycle.

The cause of this sun-spot cycle, with all the attendant changes in the general solar activity, and the source of the apparently limitless supply of solar energy still remain the two chief unsolved secrets of the sun.

## CHAPTER XVII. THE SOLAR SYSTEM.

Our sun is but a star traveling through the universe. It is accompanied in its journey to unknown parts of space, that lie in the general direction of the constellation Hercules, by an extensive family of minor bodies consisting of the eight planets and their encircling moons, one thousand or more asteroids, numerous comets, and meteors without number, all moving in prescribed paths around their ruler.

The most important members of the sun's family are the planets, Mercury, Venus, Earth, Mars, Jupiter, Saturn, Uranus and Neptune, named in the order of their position outward from the sun. We hear occasionally of the possibility of the existence of intra-Mercurial and trans-Neptunian planets and it is possible that some day an additional planet may be discovered within the orbit of Mercury or beyond the orbit of Neptune. The gravitational control of the sun extends far beyond the orbit of Neptune and there are reasons for believing in the existence of at least one or two additional planets on the outskirts of the solar system. The existence of a planet within the orbit of Mercury is now, after long continued and diligent search, believed to be very doubtful.

Were it possible to view the sun from the distance of the nearest star with the aid of the greatest telescope on earth all the members of his family would be hopelessly invisible. So, also, we cannot tell as we point our powerful telescopes at the stars whether these other suns are attended by planet families. We may only argue that it is very unlikely that there should be only one star among hundreds of millions that is attended by a group of comparatively small dark bodies that shine by the reflected light from the star they encircle.

With the exception of the two planets, Mercury and Venus, which are known as the inferior planets, since their paths lie between the earth and the sun, all the planets have moons or satellites of their own that encircle the planet just as the planet encircles the sun.

Our planet earth has one satellite, the moon, that has the distinction of being the largest of all the moons in proportion to the size of the planet it encircles. Jupiter and Saturn have moons that surpass our moon in actual size; in fact, two of the moons of the outer planets are actually larger than the smallest planet Mercury, but they are very small in proportion to the size of the planets around which they revolve. Mars, the next planet beyond the earth, the nearest of the superior or outer planets, has two tiny moons that bear the names of Deimos and Phobos, respectively. They are both less than twenty miles in diameter and revolve very near to the surface of Mars. They can only be seen with the aid of very powerful telescopes. The inner moon, Phobos, is unique in the solar system for it makes three trips around Mars while the planet is turning once on its axis.

Jupiter, the next planet outward from the sun, is almost a sun itself to its extensive family of nine moons. Four of these moons were first seen about three hundred years ago when Galileo pointed his first crude telescope at the heavens and any one can now see them with the aid of an opera glass. One of the four is equal in size to our own moon; the others surpass it in size. These moons are most interesting little bodies to observe. Their eclipses in the shadow of Jupiter, occultations or disappearances behind his disk, and the transits of the shadows as well as the moons themselves across the face of the planet can be easily seen even with the smallest telescope. The five remaining moons have all been discovered in modern times. They are extremely small bodies visible only in large telescopes. Satellite V is the nearest of all the moons to Jupiter. The other four are at great distances from the planet.

The planet Saturn has nine moons. Titan, the largest, is nearly equal in size to Jupiter's largest moon, and is larger than Mercury; four of the other moons have diameters between one thousand and two thousand miles in length. Since Saturn is nearly twice as far from the sun as Jupiter his moons are more difficult to observe, though the two largest are visible in small telescopes.

Saturn is unique in the solar system in possessing in addition to his nine satellites a most wonderful ring system, composed of swarms of minute moonlets, each pursuing its individual path around the planet. It is this unusual ring system that makes Saturn the most interesting to observe telescopically of all the planets.

The planet Uranus has four satellites and Neptune one. These planets and their satellites cannot be well observed on account of their great distances from the earth. The indistinctness of surface markings makes it impossible to determine the period of rotation of these two outer planets on their axes. It is believed that their rotation is very rapid, however, as is the case with the other planets Jupiter and Saturn.

All the planets in the solar system fall naturally into two groups. Jupiter, Saturn, Uranus, and Neptune, the members of the outer group, have on the average, diameters ten times as great and, therefore, volumes one thousand times as great as Mercury, Venus, Earth and Mars, the members of the inner or terrestrial group.

The terrestrial planets are the pigmies of the solar system, the outer planets are the giants. The densities of the planets Mercury, Venus, Earth and Mars are several times greater than the density of water. They are all extremely heavy bodies for their size, and probably have rigid interiors with surface crusts.

The existence of life on Mercury is made impossible by the absence of an atmosphere. Venus and Mars both have atmospheres and it is possible that both of these planets may support life. Mars has probably been the most discussed of all the planets, though Venus is the Earth's twin planet in size, mass, density and surface gravity, just as Uranus and Neptune are the twins of the outer group. It is now believed that water and vegetation exist on Mars. The reddish color of this planet is supposed to be due to its extensive desert tracts. The nature of certain peculiar markings on this planet, known as canals, still continues to be a matter of dispute. It is generally believed since air, water and vegetation exist on Mars, that some form of animal life also exists there.

The length of the day on Mars is known very accurately, for the rareness of its atmosphere permits us to see readily many of its surface markings. The length of the day is about twenty-four and one-half hours, and the seasonal changes on Mars strongly resemble our own, though the seasons on Mars are twice as long as they are on our own planet since the Martian year is twice as long as the terrestrial year.

The question of life on Venus depends largely upon the length of the planet's rotation period. This is still uncertain since no definite surface markings can be found on the planet by which the period of its rotation can be determined. So dense is the atmosphere of Venus that its surface is, apparently, always hidden from view beneath a canopy of clouds. It is the more general belief that Venus, as well as Mercury, rotates on its axis in the same time that it takes to make a revolution around the sun. In this case the same side of the planet is always turned toward the sun and, as a result, the surface is divided into two hemispheres—one of perpetual day, the other of perpetual night.

This peculiar form of rotation in which the period of rotation and revolution are equal is by no means unknown in the solar system. Our own moon always keeps the same face turned toward the earth and there are reasons for believing that some of the satellites of Jupiter and Saturn rotate in the same manner.

Life on any one of the outer planets is impossible. The density of these planets averages about the same as the density of the sun, which is a little higher than the density of water. The density of Saturn is even less than water. In other words, Saturn would float in water and it is the lightest of all the planets. It is assumed from these facts that the four outer planets are largely in a gaseous condition. They all possess dense atmospheres and, in spite of their huge size, rotate on their axis with great rapidity. The two whose rotation periods are known, Jupiter and Saturn, turn on their axis in about ten hours. On account of this rapid rotation and their gaseous condition both Jupiter and Saturn are noticeably flattened at the poles.

The terrestrial planets are separated from the outer group by a wide gap. Within this space are to be found the asteroid or planetoid group. There are known to be over nine hundred and fifty of these minor bodies whose diameters range from five hundred miles for the largest to three or four miles for the smallest. There are only four asteroids whose diameters exceed one hundred miles and the majority have diameters of less than twenty miles. The total mass of the asteroids is much less than that of the smallest of the planets. It was believed at one time that these small bodies were fragments of a shattered planet, but this view is no longer held. The asteroids as well as the comets and meteors probably represent the material of the primitive solar nebula that was not swept up when the larger planets were formed.

With few exceptions the asteroids are only to be seen in large telescopes and then only as star-like points of light. Most of them are simply huge rocks and all are necessarily devoid of life since such small bodies have not sufficient gravitational force to hold an atmosphere.

The revolution of the planets around the sun and of the satellites of the planets around the primary planets are performed according to known laws of motion that make it possible to foretell the positions of these bodies years in advance. Asteroids and comets also obey these same laws, and after three observations of the positions of one of these bodies have been obtained its future movements can be predicted. All the planets and their satellites are nearly perfect spheres. They all, with few exceptions, rotate on their axes and revolve around the sun, or, in the case of moons, around their primaries, in the same direction, from west to east. Only the two outermost satellites of Jupiter, the outermost satellites of Saturn and the satellites of Uranus and Neptune retrograde or travel in their orbits from east to west, which is opposite to the direction of motion of all the other planets and satellites.

The paths of all the planets around the sun are ellipses that are nearly circular, and they all lie nearly in the same plane. The asteroids have orbits that are more flattened or elliptical and these orbits are in some instances highly inclined to the planetary orbits. The comets have orbits that are usually very elongated ellipses or parabolas. Some of the comets may be only chance visitors to our solar system, though astronomers generally believe that they are all permanent members. Paths of comets pass around the sun at all angles and some comets move in their orbits from west to east, while others move in the opposite direction or retrograde. The behavior of the asteroids and comets is not at all in accord with the theory that was, until recently, universally advanced to explain the origin of the solar system.

Some astronomers have made attempts to modify the nebular hypothesis that has held sway for so many years, in order to make it fit in with more recent discoveries, but others feel that a new theory is now required to explain the origin of the solar system. Several theories have been advanced but no new theory has yet definitely replaced the famous nebular hypothesis of the noted French astronomer La Place.

## CHAPTER XVIII. THE ORIGIN OF THE EARTH.

It is not possible to consider the question of the origin of the earth apart from the question of the origin of the solar system. That all the planets, as well as the asteroids, originated from a common parent-mass has never been seriously questioned. All of these bodies revolve about the sun, and rotate upon their axes in the same direction—from west to east. Moreover, all of the planetary orbits lie very nearly in the same plane and are nearly circular in form.

The orbits of the asteroids are more elliptical and more highly inclined to one another than are the orbits of the planets, but on the average they are neither very elliptical nor very highly inclined to the planetary orbits.

The sun rotates upon its axis in the same direction in which the planets rotate and perform their revolutions, and the orbits of the planets are inclined at small angles to the plane of the sun's equator.

These facts are all significant and cannot be overlooked in formulating a theory to explain the origin of the planetary system in general and of the earth in particular. Presumably the planets and asteroids formed at one time a part of a central body which rotated on its axis in the direction in which they now revolve about the sun.

When and by the operation of what force, external or internal, they were separated from this central body is the question.

In 1796 La Place advanced his celebrated nebular hypothesis to explain the origin of the solar system. It was received with favor both by scientists and laymen, and in a short time was almost universally accepted as closely approximating to the truth.

According to the nebular hypothesis the solar nebula from which the planetary system was formed, originally extended at least as far as the orbit of Neptune and rotated slowly in the direction in which the planets now revolve. As it lost heat by radiation and contracted under the gravitation of its parts its rate of rotation increased. When the centrifugal (center-fleeing) force at the equator equalled the gravitational force directed toward the center, a ring would be left behind by the contracting nebula. Such a ring would not be absolutely uniform and would break at some point and gather into a planetary mass under the gravitation of its parts. This planetary mass would abandon rings in turn and these would break up to form satellites. Successive rings were supposed to have been abandoned at intervals by the solar nebula at the present distances of the planets from the sun in the manner described above until the original solar nebula had contracted to its present size.

The rings of Saturn were supposed to be the single example remaining of this process of forming planets and satellites from a contracting nebulous mass.

The La Placian hypothesis attempted to explain why all the planets and their satellites revolve in the same direction in which the sun turns on its axis, in nearly circular orbits and nearly in the same plane. At the time it was advanced it appeared to be in accord with all the facts then known regarding the solar system.

The planetoids with their interlacing and in some instances highly inclined and elliptical orbits were then undiscovered. It would have been impossible for them to have been formed by the abandonment of successive rings from a central, rotating mass.

The constitution of Saturn's rings was unknown at this time; also the fact that the moonlets of the inner ring revolve about Saturn in half the time required for the planet to turn on its axis—another impossibility under the nebular hypothesis, for, according to the assumptions of the nebular hypothesis it would be impossible for a satellite to revolve about a central body in a shorter time than that body turns on its axis.

The satellites of Mars were not discovered until many years later, as well as the retrograding satellites of Jupiter and Saturn, all presenting difficulties in the way of accepting the nebular hypothesis without radical changes. Attempts, mostly unsuccessful, have been made from time to time to make these exceptional cases fit in with the requirements of the nebular hypothesis.

The theory that the sun's heat was maintained by the contraction of the original solar nebula, which would cause its temperature to rise, appeared to give considerable support to the theory of La Place, but the mathematicians got to work and showed that the amount of heat that would be furnished by the contraction of the sun from beyond the orbit of Neptune to its present dimensions would be sufficient to supply heat to the earth at the present rate for only twenty-five million years, a period far too brief, the geologists and biologists said, to cover all the vast cyclical changes that are known to have taken place upon the surface of this planet since its surface crust was formed. Evidently gravitational contraction is by no means the only or even the chief source of the sun's heat.

It was also shown indisputably, that it would have been impossible for successive rings to have been abandoned at certain definite intervals by a contracting nebula, and granted a ring could have been formed it would have been impossible for it to condense into a planet, since forces residing in the sun would offset the gravitation of its parts.

When La Place advanced his famous theory it was, to use his own words, "with that distrust which everything ought to inspire that is not a result of observation or of calculation."

Were La Place living today he would be, we believe, the first to abandon a theory that is now known to be in accord neither with observation nor calculation.

Deprived of a theory that has served to explain the outstanding features of the solar system more or less adequately for one hundred and twenty-five years, astronomers are seeking in the light of recent observations and discoveries to formulate a satisfactory theory of the origin of the solar system.

In the planetesimal theory of Chamberlin and Moulton and the recently suggested theory of the well-known English mathematician, Jeans, a second sun passing close to our own sun is assumed to have been the cause of the origin of the planetary system.

The effect of the close approach of such a sun would be the ejection of a stream of matter from our sun, as we term it, in the direction of the passing body and also in a diametrically opposite direction. This ejection would be continuous as long as the stars remained near one another, the height attained by the ejected stream decreasing as the passing star receded. The result would be the formation of a spiral nebula in which the motion of the ejected particles—planetesimals—would be across the spiral arms, toward and away from the passing star. After the sun had receded so far as to have no further effect upon these ejected particles they would revolve about the sun in more or less elliptical orbits which would gradually be reduced to nearly circular forms by repeated collisions between planetesimals. Larger nuclei would be formed and these would gradually sweep up smaller fragments and become the planets of the present system. Smaller nuclei in the vicinity of larger ones would become their satellites and in the course of many millions of years all of the larger fragments would be swept up by the planetary nuclei and their satellites—leaving only the asteroids, comets and meteors as survivors of the original spiral system.

It must be borne in mind that a spiral nebula formed by the close approach of two suns would resemble in form only the great spiral nebulæ that are known to exist by hundreds of thousands in the heavens. These are far too extensive to form anything so small as a single solar system, but might condense into systems composed of many suns—either galaxies or star clusters.

Jean's suggested theory of the origin of the planetary system differs in its details from the above, though a passing sun is assumed to be the disturbing force that causes the ejection of a stream of matter which condenses to form the planets and their satellites. The origin of the inner planets is left greatly in doubt by this theory, however, and the system which interests us chiefly—the earth-moon system—is the one about which it is most difficult to arrive at any definite conclusion. Our own sun, it is assumed, was dark and cold, of low density and with a diameter about equal to that of Neptune's orbit at the time of the catastrophe which is placed at some 300,000,000 years ago. In Jean's words, "... The time for arriving at conclusions in cosmogony has not yet come—and it must be left to future investigators armed with more mathematical and observational knowledge than we at present possess to pronounce a final decision."

However, since La Place advanced his celebrated nebular hypothesis, great advances in astronomy have been made, and man is in a better position to theorize on this fascinating problem today than he was one hundred and twenty-five years ago.

All such theories must necessarily be regarded as working hypotheses only, to be discarded or modified as our knowledge and understanding of the laws of the universe increase. No theory can ever be regarded as final or perfect.

The discovery of radio-activity furnishes us with new material for new theories. The sun and the planets may be and probably are far older than we ever dreamed could be possible. It is no longer necessary or reasonable to assume that a greatly extended solar nebula once existed and supplied the planets with heat through gravitational contraction or to place a time limit upon the period required for the formation of the planets and their satellites that is not in accord with the requirements of other sciences.

We know today that there exist within the sun powerful repulsive forces, which even under present conditions occasionally eject gaseous matter to heights of five hundred thousand miles or more with a velocity of over two hundred miles per second. Small changes in the velocity of ejection produce great differences in the height of the ejected columns.

With an initial velocity of three hundred and eighty miles per second, matter would be thrown from the solar surface to a height of fifty million miles. Were the velocity of ejection three hundred and eighty-three miles per second the height of the column would be five hundred million miles, while a further increase in the initial velocity would send matter away from the sun, never to return.

Instead of suns and solar systems evolved from nebulæ we are now more familiar with the idea of nebulæ evolved from stars through some terrific cataclysm as in the case of novas or temporary stars.

It is now known that there exist in certain parts of space a number of sharply defined stars surrounded by extensive nebulous envelopes. Are these possibly suns that are going through the process of forming their planetary systems?

It is now known that pressure of light and electrical repulsion are forces to be reckoned with in the evolution of stars and nebulæ as well as gravitational contraction. It has long been felt that the peculiar formations existing among the vast irregular gaseous nebulæ could not be explained as gravitational effects alone.

Light-pressure and electrical repulsion, as well as gravitation are at work within the solar system and the sun is the seat of powerful disturbances which produce periodic outbursts of exceptional activity and which may have produced in the distant past more startling effects than any with which we are familiar at present.

The earth and moon form a system that is in a way unique. No satellite in the solar system is so large in proportion to its primary as is our own moon. Seen from the distance of Venus or Mars, the two bodies would apparently form a double star. The diameter of the moon is one-fourth that of the earth. Satellite III of Jupiter far exceeds our own moon in actual size but its diameter is only about four-hundredths of the diameter of the planet about which it revolves. The diameter of Titan, the largest satellite of the Saturnian system, bears the same ratio to the diameter of Saturn. Moreover, all the nearer satellites of Jupiter and Saturn lie nearly in the equatorial planes of these planets, but the orbit of the moon is inclined at a high angle to the plane of the earth's equator.

It is not difficult to believe that the satellites of Jupiter or Saturn were at some time thrown off from the equatorial belts of their primaries, just as the planets themselves may have been ejected from the equatorial belt of the sun, but we cannot so readily believe that our own satellite was formed from the earth in a similar manner.

The moon's orbit lies nearly in the plane of the sun's equator, however, and it is conceivable that both earth and moon were simultaneously ejected from the equatorial zone of the sun, the two nuclei being so close together that the smaller one remained under the gravitational control of the larger.

The difficulties in the way of believing that the moon once formed a part of the earth are very great. It can be shown mathematically that if the two bodies at one time formed a single mass it would have been impossible for the moon to break away from the earth, unless the force that caused the separation were sufficient to hurl the moon to a greater distance than two and a half times the earth's radius. The mathematician, Roche, found out by computation that a satellite could not remain intact within this distance of the planet, but would be broken up into small fragments under the effects of the tides raised by the larger body. If, then, the moon had originally been ejected from the earth to a less distance than two and one-half radii of the earth (2.44 to be exact) it would have been disintegrated into small particles, or moonlets, under the tidal strains exerted upon it by the earth and would have been gradually distributed about the earth in the form of a meteoric ring which, in the course of ages, would be absorbed by the earth, just as Saturn is now gradually absorbing its rings.

The planets differ greatly in density. The more distant and larger planets—Jupiter, Saturn, Uranus and Neptune—have densities equal to or less than that of the sun. The densities of the inner planets—Mercury, Venus, Earth and Mars—are, relatively, extremely high, the density of the Earth's core being about that of meteoric iron. The densities of Mercury and Venus are slightly less than that of the earth and the densities of Mars and the moon about equal to that of the earth's crust.

If a stream of matter were ejected from the sun under the influence of some external force, such as that exerted by a passing star, the outlying parts of the stream would consist of the lighter elements and the lower parts of the heavier elements, since the lighter solar elements lie at or near the surface of the sun and the heavier elements at greater depths. At the time of ejection the lighter elements would be thrown to great distances and would go to form the less dense outer planets; the heavier elements would go to form the inner planets of high density.

It is conceivable that ejection of solar material might have taken place under the influence of certain forces at work within the sun itself, such as electrical repulsion or pressure of light which might become powerful enough under certain conditions to overcome the effect of gravitation.

Next to nothing is known about the physical state of matter at great solar depths, where abnormal conditions of temperature and pressure must exist, and where great physical changes and disturbances may have taken place in the past. Even today solar activity goes through a cycle of change during the sun-spot period, and many millions of years ago the sun-spot cycle of solar activity may have been far different from what it is today and a far more powerful factor in producing changes in the solar system.

Outbursts of novas indicate that agencies making for peace and order are not the only ones at work among the stars. The cause of such outbursts has never been satisfactorily explained. The theory that they are caused by the close approach of two suns or by the encounter of a star with a dark nebula does not account for all of the circumstances of such outbursts. The nebulous matter seen about a nova after the outburst is now generally believed to have been expelled from the star itself at the time of the catastrophe and may conceivably be the stuff of which planetary systems are made.

At some epoch in the past, probably at least one thousand million years ago, our own sun may have undergone some cataclysmic change and this may, conceivably, have been brought about by disturbances within the sun itself. Elements may have been so formed and distributed within the interior of the sun that friction and internal instability resulted and in time produced an upheaval of solar elements with initial velocities so great that, possibly, through electrical repulsion and light-pressure, portions of the ejected streams were permanently detached from the sun and became the nuclei of future planets. In some such way, it is conceivable, our own planet Earth and the other members of our solar system may have been brought into existence in the dim and distant past—many hundred million years ago.

## CHAPTER XIX. JUPITER AND HIS NINE MOONS.

Jupiter shines by reflected sunlight with a brilliancy that usually exceeds that of the brightest of the stars, Sirius. When seen during the midnight hours the remarkable unflickering brightness of this largest and most distinguished member of the solar system at once serves to set it apart from the scintillating stars far beyond.

There is but one planet, Venus, that always surpasses Jupiter in brilliancy, though Mars on the occasions of its close approaches to the earth may equal or slightly surpass Jupiter in brightness. As Venus never departs more than forty-eight degrees from the sun, and so is never seen in the midnight hours, Jupiter usually shines without a rival when visible at midnight. To one who has observed the two planets together the silvery radiance and surpassing brilliancy of Venus, due not to its size, but to its comparative nearness to the earth, at once serves to distinguish it from the golden glow of Jupiter.

Even the smallest telescopes of two- or three-inch aperture will show the four historic moons of Jupiter which were the first celestial objects to be discovered when Galileo turned his crude telescope to the heavens in the year 1610.

The fact that these tiny points of light were actually revolving around the great planet was soon detected by the famous astronomer and we can imagine with what breathless interest he observed these satellites of another world whose discovery dealt such a severe blow to the old Ptolemaic theory that the earth was the center of the universe. It was not until the great telescopes of modern times were invented that the five additional moons of Jupiter were discovered. The four satellites first observed by Galileo were fancifully named Io, Europa, Ganymede and Callisto, in the order of their positions outward from the planet, but these names are rarely used now, the satellites being designated for convenience I, II, III and IV, respectively. The first of the new satellites to be discovered was Satellite V, which is the nearest to Jupiter of all the nine moons. It is an extremely small body, not more than one hundred miles in diameter, and to discover this tiny body as it skirted rapidly around the great planet within sixty-seven thousand miles of its surface, nearly lost in the glaring rays, was a difficult feat even for an experienced observer. It was accomplished, however, by Prof. E. E. Barnard with the great Lick refractor in 1892. Satellite V is hopelessly beyond the reach of any but the greatest telescopes, as are also the four satellites discovered since that date. In fact, most of these tiny moons are observed photographically. Satellites VI and VII were discovered photographically in 1905. They are both about seven million miles from the planet and their paths loop through one another; they are, moreover, highly inclined to each other at an angle of nearly thirty degrees. When nearest together they are separated by a distance of two million miles. Two more extremely small bodies, known as Satellites VIII and IX, have been discovered since then, one at Greenwich, England, in 1908, the other at the Lick Observatory in 1914. These excessively faint bodies are the most remote satellites of Jupiter and they are of particular interest because they travel around the planet in a retrograde direction, or from east to west, which is opposite to the direction of revolution prevailing in the solar system. The ninth and most distant satellite of Saturn also retrogrades, or revolves in a clockwise rather than a counter-clockwise direction around the planet. One explanation given for this peculiarity of the outermost satellites of Jupiter and Saturn is that this backward revolution around the planet is more stable when the satellites are at great distances from the primary, and the gravitational control that the planet exerts is therefore weak. The moons of the planets are, of course, subject to the attraction of the sun as well as to the attraction of the controlling planet, and the greater the distance of the satellite from the planet the stronger the pull exerted by the sun and the weaker the bonds that bind the moon to the planet. Beyond a certain limit it would be impossible for the planet to hold the satellite against the sun's greater attraction and the satellite would leave the planet to revolve directly around the sun, thereby becoming a planet. It appears that as this danger limit is neared it is safer for the satellite to "back" around the planet than to follow the usual "west to east" direction of revolution. The eighth satellite of Jupiter is more than fourteen million and the ninth more than fifteen million miles from the parent planet and they require about two years and three years, respectively, to complete one trip around Jupiter. When we consider that Satellite V darts around the planet in less than twelve hours at a distance of only sixty-seven thousand miles from its surface we realise what tremendous differences exist in the distances and periods of revolution of the nine moons. There is also great disparity in the sizes of the various moons. The five moons discovered in modern times are all excessively faint and extremely small. The diameter of the largest of these, Satellite V, is less than one hundred miles. On the other hand, the four historic moons of Jupiter are of planetary dimensions. The smallest, Satellite II, is slightly larger than our own moon, while the largest, Satellite III, has a diameter, according to measurements made with the 40-inch Yerkes refractor in 1916, of three thousand nine hundred and eight miles, which is only four hundred miles less than the diameter of Mars. The periods of revolution of these four satellites range from one day and eighteen hours for the nearest, which is about two hundred and sixty-one thousand miles from the center of Jupiter, to sixteen days and sixteen and one-half hours for the most distant, which is more than one million one hundred and sixty thousand miles from the planet. These four moons are so near to the great planet that they are continually dipping into his huge shadow and experiencing an eclipse of the sun which, owing to the nearness and great size of Jupiter, lasts for two or three hours. At times of eclipse the moon suddenly disappears from the observer's view, though it may be considerably to one side of the planet. Its reappearance later on is just as sudden, or it may pass out of the shadow while hidden from us behind the disk of the planet, in which case its reappearance is invisible from the earth. The occultations of the satellites, or, in other words, their disappearance behind the planet's disk, are also interesting phenomena to observe, as are their "transits" across the disk of the planet as the satellite passes in front of it. Not only the satellite itself but its shadow as well can be seen, a small black dot passing over the surface of Jupiter. The satellite is totally eclipsing the sun for this small dark portion of the planet's disk. Two satellites and their shadows are frequently seen crossing the face of the planet at the same time. It is possible to observe all the phenomena of the satellite's transits and shadows, eclipses and occultations with very small telescopes. From observations of the eclipses of Jupiter's satellites the important discovery of the finite velocity of light was first made as far back as the year 1675.

Faint surface markings have been made out at certain times on the largest of the four satellites, Satellite III, and also on Satellite I. Observations of the markings on the former seem to indicate that it always keeps the same face turned toward Jupiter as does our own moon toward the earth.

There are also reasons for believing that the equatorial regions of Satellite I are light colored and the polar regions dark. There is the possibility that forms of life may exist on these satellites of Jupiter, though they are more likely barren, lifeless worlds, such as Mercury and the moon. Their great distance from the earth, never less than three hundred and sixty-eight million miles, makes observations of their surface markings very difficult.

How beautiful beyond description must the heavens appear as viewed from the satellites of Jupiter! Viewed from the distance of Io, or Satellite I, the mighty planet Jupiter presents a spectacle such as the eye of man has never been privileged to behold. The huge flattened globe, ninety thousand miles in equatorial diameter, equal in mass to three hundred planets such as our own and in volume to nearly fourteen hundred, fills a space in the heavens nearly twenty degrees in extent as viewed from this satellite. Fifteen hundred of our own full moons would hardly fill the same space. Whirling on its axis with frightful speed in a period of less than ten hours, the huge ball glides rapidly but majestically onward through the sky. A far distant sun shrunk to but one-fifth the diameter of the full moon throws light and shade across the rapidly changing surface of the planet, rich in the reds, browns and yellows and all the gorgeous shades and tints of its dense, seething, gaseous envelope. The phases of the moon on a greatly enlarged scale rapidly succeed each other on Jupiter as it is viewed from the satellite in all positions with reference to the sun. The cause of the belts of Jupiter, that lie parallel to the planet's equator and are constantly changing in number, width and shade, as well as the nature of all the peculiar splashes of color and intensely white flecks that come and go in the dense atmosphere of the planet would not be such a mystery to us were it possible to view the great planet from the distance of Satellite I, which is about as far from the surface of Jupiter as the moon is from the earth. It is uncertain whether the planet is entirely gaseous throughout or has a central core of solid or liquid matter. Its density is only one and one-quarter times that of water and slightly less than that of the sun, showing that it is composed largely, if not entirely, of matter in a gaseous state. Jupiter is a world as different from our own as it is possible to imagine. There is no visible surface crust and there are no permanent markings. Different spots on the planet's disk give different periods of rotation showing that it is atmospheric phenomena that we observe. All is constant flux and change on Jupiter. Dense vapors arise from a highly heated interior and spread out into belts parallel to the equator in the direction of the planet's rotation. From its nearest satellite all the interesting changes of color and form that constantly take place in the atmosphere of this great globe could be observed in great detail. The high percentage of light and heat that Jupiter reflects from the sun to its nearer satellites makes it a secondary sun to them of tremendous size though feeble strength.

As seen from Satellite I the other three major moons of Jupiter present all the phases of our own moon in rapid succession, due to their constantly changing positions with reference to the sun. The five small moons, discovered in modern times, are so minute that they are simply star-like points of light even when viewed from the other moons of Jupiter.

To keep track of the rapidly changing positions and various phases of the moons of Jupiter as seen from any one of them, as well as the rapid apparent motion of the planet through the sky due to the revolutions of the satellite around the planet, would be a troublesome task for an astronomer stationed on one of these far distant worlds. It would be a common sight to see in the sky at one time the huge planet, the far-distant, shrunken sun, and one, two or three moons. Seen from the moons of Jupiter the constellations would appear as they do to us on earth, for such a slight change in position as five hundred million miles, more or less, is trivial when one is looking at the stars. Observations of the stars from the nearest moon of Jupiter would be attended with great difficulties at times, since reflected sunlight from a body nearly twenty degrees in diameter would be extremely troublesome, especially were the phases of the planets near that of the full moon. We know how the presence of our own moon in the heavens at the full dims the brightness of the stars so that only the brightest stars are seen. Even as viewed from the fourth or most distant of the major satellites the planet subtends an angle of nearly five degrees. Occultations of the stars are many and frequent as the huge planet globe glides swiftly through the heavens. Many a moonlight night appears almost as day owing to the presence of the enormous, brilliantly reflecting ball of light and at times two or three moons in addition. Only the brightest stars could possibly be seen under such circumstances. When, however, the small worlds pass into the shadow of the great mother planet and not only the light of the sun but also the reflected light of Jupiter disappears for many minutes, the stars shine forth in all their glory there as here. At such times some of the larger moons would usually be seen shining by the reflected light of the far distant sun. Saturn also would be visible as a magnificent star, but beautiful Venus and ruddy Mars would fail to appear. Tiny bodies, mere specks of light at this distance, they would be lost to view in the glare of the sun.

## CHAPTER XX. THE RINGS AND MOONS OF SATURN.

Nearly everyone has felt at some time or other a strong desire to gaze at some of the beauties and wonders of the heavens through a telescope and the one object that all of us wish to see, if, perchance, this desire is to be gratified is Saturn, whose unusual ring system has, so far as we know, no counterpart in the sky.

All the planets in the solar system with the exception of the two innermost, Mercury and Venus, are attended by satellites, but Saturn, alone, has in addition to a family of nine moons, three distinct rings of great dimensions which are composed of swarms of minute particles revolving around the planet.

Why Saturn should be the only planet to possess such a system of rings has never been explained in an entirely satisfactory manner. There is an interesting law known as "Roche's Law," however, named from its investigator, that states that no satellite of a planet can exist intact with 2.44 times the radius of the planet. This limit is spoken of as "Roche's Limit" and applying it to the planet Saturn we find that the rings of Saturn fall within this limit. It does not necessarily follow from this that the minute particles of which the rings are composed are the shattered remains of one small satellite, but rather that they are the material from which a satellite might have been formed were it not so close to the planet. Within "Roche's Limit" the mutual attraction of the various particles for each other that would tend eventually to gather them into one body is overcome by tidal forces that arise from such close proximity to the huge planet. The stress and strain of such forces is so great that no grouping of particles can take place. This explains, possibly, why the rings continue to exist in their present condition. The total quantity of matter in the rings is known to be very small, for it does not disturb the motions of any of the nearer and smaller satellites, though tiny Mimas, six hundred miles in diameter, is only thirty-one thousand miles beyond the outer edge of the outer ring.

An interesting observation was made a few years ago of the passage of the rings of the planet between us and a star. Though the light of the star was diminished to one-fourth of its normal brightness when the rings passed before it, at no time was its light entirely eclipsed by any of the particles. It was computed that if the diameters of any of the individual particles had amounted to as much as three or four miles the star would have been temporarily eclipsed. An upper limit for the size of the moonlets was thus obtained. The average diameter of the particles is probably much less than three miles.

The thickness of the ring system is not over fifty or one hundred miles, but its total diameter is one hundred and seventy-two thousand miles. There are, in all, three concentric rings. The faint inner ring, known as the "crape" ring, is invisible in a telescope under four inches in aperture. The width of this inner ring is eleven thousand miles. Just beyond the crape ring is the chief, bright ring, eighteen thousand miles in width. It shades gradually in brightness from its juncture with the crape ring to its most luminous portion at its outer edge, which is separated from the third or outer ring by a gap two thousand two hundred miles in width, known as Cassini's Division. The third or outer ring is eleven thousand miles wide and is less bright than the central ring. The inner edge of the inner ring is but six thousand miles above the surface of the planet. On account of the curvature of the planet the ring system is invisible from the north and south poles of Saturn. As in the case of the satellites of a planet the inner particles of the rings revolve around the planet more rapidly than the outer particles. The innermost particles of the crape ring require but five hours for one journey around Saturn while the outermost particles of the outer ring require one hundred and thirty-seven hours, or nearly six days to complete one revolution.

In addition to the gap in the rings known as Cassini's Division several other fainter divisions exist. If a group of moonlets were to revolve around the planet in the positions marked by these gaps their periods of revolution would be commensurable with the periods of several of the satellites of Saturn. As a result the attraction exerted on such particles by these satellites would gradually disturb their motion in such a way as to draw them away from these positions. It is owing, therefore, to the attraction of the satellites of Saturn for the moonlets that these gaps in the rings exist.

As a result of the disturbances produced in the motion of the moonlets by the satellites of Saturn collisions are bound to occur occasionally among the various particles. When two particles collide the period of revolution of one or both of them is reduced and as a result collisions tend to bring the moonlets gradually closer and closer to the surface of the planet. The dusky inner ring, it is believed, may consist largely of particles whose periods have been continually shortened by collisions.

Saturn may, therefore, lose its ring system in the course of time through its gradually being drawn down upon the planet by collisions of the various particles until all of the material is finally swept up by the planet. Such a change would probably require millions of years, however, as collisions are probably, on the whole, infrequent. It is possible that the ring system of Saturn may have been much more extensive in the past than it is now and other members of our solar system may have had such appendages in the far distant past.

The appearance of the rings of Saturn as viewed from our planet changes periodically as a result of the revolution of the earth and Saturn around the sun, which places them in constantly changing positions with reference to each other. The rings lie in the plane of Saturn's equator, which is inclined twenty-seven degrees to its orbit and twenty-eight degrees to the Earth's orbit.

Since the position of the equator remains parallel to itself while the planet is journeying around the sun it happens that half the time the earth is elevated above the plane of the rings and the remainder of the time it lies below the plane of the rings. Twice in the period of Saturn's revolution around the sun, which occupies nearly thirty years, the earth lies directly in the plane of the rings and at this time the rings entirely disappear from view for a short time. Mid-way between the two dates of disappearance the rings are tilted at their widest angle with reference to the earth and they are then seen to the best advantage. As the date of their disappearance approaches they appear more and more like a line of light extending to either side of the planet's equator. Even in the most powerful telescope the rings entirely disappear from view for a few hours at the time the earth lies exactly in the same plane. It is at this time that the ball of the planet is best seen. Its flattening at the poles, which is nearly ten per cent. of its equatorial diameter then gives it a decidedly oval appearance. Ordinarily one of the hemispheres of Saturn is partly or entirely concealed by the rings so that the oblate form is not so noticeable. It was the change in the tilt and visibility of the rings that so perplexed Galileo when he attempted to make out the nature of these appendages of Saturn with his crude telescope of insufficient magnifying power. So great was his bewilderment when the rings finally disappeared that he cried out in despair that Saturn must have swallowed his children, according to the legend. He finally became so exasperated with the results of his observations that he gave up observing the planet. The true nature of these appendages of Saturn remained a mystery until Huygens solved the problem in 1655, some time after the death of Galileo.

In addition to the rings, Saturn has nine satellites named, in the order of their distance outward from the planet, Mimas, Enceladus, Tethys, Dione, Rhea, Titan, Hyperion, Iapetus and Phoebe. The last-mentioned satellite was discovered by W. H. Pickering in 1899. It aroused great interest at the time because it was the first satellite to be discovered with "retrograde" motion in its orbit. Two satellites of Jupiter since discovered revolve in the same direction around their primary.

The satellites of Saturn are approximate to those of Jupiter in size and exactly equal them in number. The largest, Titan, is three thousand miles in diameter and can be easily seen with the smallest telescopes. With a four-inch telescope five of the satellites can be readily found, though they are not as interesting to observe as the satellites of Jupiter because they are far more distant from the earth. The time they require to make one journey around Saturn varies from nearly twenty-three hours for Mimas, the nearest, to approximately five hundred and twenty-four days for Phoebe, the most distant.

Saturn as well as Jupiter is marked by belts parallel to the Equator though they appear more indistinct than those of Jupiter on account of the greater distance of Saturn. Saturn also resembles Jupiter in its physical composition which is largely, if not entirely, gaseous, and in the extremely short period of rotation on its axis which is approximately ten hours. In more ways than one Saturn is a very unusual planet. In addition to possessing an enormous ring system it is the lightest of all the planets, its density being only sixty-three hundredths that of water, and it is the most oblate, its flattening at the poles amounting nearly to one-tenth of its diameter. Its equator is more highly inclined to its orbit than is the case with any other planet, not even excepting the earth and Mars. For this reason its seasonal changes are very great, in marked contrast to Jupiter whose equator lies very nearly in the plane of its orbit. Since Saturn is so far away from the sun that it receives only one ninetieth as much light and heat per unit area as the earth, its outer gaseous surface must be extremely cold unless considerable heat is conveyed to the surface from within its hot interior.

The late Prof. Lowell concluded from certain observations made at Flagstaff, Ariz., that Saturn is composed of layers of different densities and that the inner layers are more flattened at the poles and rotate faster than the outer layers. Marked variations in the color and brightness of the ball of the planet have been noted from time to time. In 1916 observers of Saturn described the planet as pinkish-brown and conspicuously darker than the brighter portions of the rings.

It is believed that these very noticeable changes in the color and brightness of Saturn are due to slight, irregular changes in the intensity of the radiations of the sun which set up certain secondary effects in the atmosphere of the planets. Similar changes in color and brightness have been observed also in the case of Jupiter.

# CHAPTER XXI. IS THE MOON A DEAD WORLD?

It has been a generally accepted belief among astronomers for years that the moon is a dead world devoid of air and water and so, necessarily, lifeless. It is certain that the moon has no extensive atmosphere such as envelops our own planet. There is abundant proof of this fact. The edge of the lunar disk is clear-cut. Whenever, as happens frequently, the moon passes between us and a star the disappearance of the star is instantaneous. There is no gradual dimming or refraction of the star's light by atmospheric vapors. Moreover, lunar shadows are harsh and black. There is no evidence of diffusion of light on the moon by atmospheric gases.

The absence of water or water vapor on the visible surface of the moon, at least in any appreciable quantity, is plainly evident to anyone who observes the moon through the telescope. Even with small telescopes, objects five miles or so in diameter can be readily detected and clouds drifting over the surface could not possibly escape our observation if they existed.

Bodies of water, great or small, would be plainly visible and would besides give rise to water vapor and clouds, which we would not fail to detect.

Since the surface of the moon is unscreened by air and water vapor to absorb the incoming rays from the sun, and the outgoing radiations from the surface, the extremes of temperature between day and night are very great, and are augmented by the fact that the lunar day equals the lunar month in length, so that fourteen days of untempered heat are followed by fourteen days of frigid darkness. Observations of the rate of radiation from the moon's surface during total eclipses of the moon indicate that the moon's radiation is very rapid, and that its temperature during the height of the lunar day probably approaches 200° F., while at the lunar midnight it may have fallen to 100° below zero F., or even lower.

With air and water both lacking and such extremes of temperature existing why should we seriously consider the question of life on the moon?

This is the point of view of the majority of astronomers and it seems well taken. Yet many astronomers who have made a special study of the lunar surface for years under all conditions of illumination and phase, and have most carefully observed and mapped and photographed its characteristic markings, are agreed that there are evidences that changes are taking place on the moon, and recently Prof. W. H.

Pickering has expressed the belief, substantiated by drawings, that there is a progressive change of color or darkening within certain lunar craters with the advance of the lunar day, indicating, in his opinion, a rapid vegetational growth that springs up in the height of the lunar day and dies out as the lunar night approaches.

Some years ago certain selenographers suggested that there might exist in the numberless crater-pits and craters, in the deep-lying maria or "seas," and in the clefts and rills and cracks that form intricate systems all over the lunar surface, certain exhalations from the surface and heavy vapors including possibly carbon dioxide and water vapor to temper the extremes of the long lunar days and night and furnish the necessary medium for the support of certain forms of animal and vegetable life.

Many astronomers, including a number who are not in sympathy with the above view, believe that snow and ice exist on the moon, even though water in the form of liquid and vapor is not observable. All the extremely brilliant portions of the surface, according to some astronomers, are covered with snow and ice. Certainly, some portions of the moon's surface reflect sunlight as brilliantly as if they were covered with freshly fallen snow, while other portions appear to be black by contrast. There also appears to be evidence that certain small markings, described as crater-cones and resembling our terrestrial volcanoes more than any other lunar feature, are at times temporarily obscured from view by a veil of vapors. Many observers believe that these crater-cones are active volcanic vents, and that there is considerable volcanic activity still taking place upon the moon.

These small crater-cones resemble, we are told, parasitic cones found on the sides of terrestrial volcanoes, and they are frequently seen on the floors of craters closely associated with light streaks. These crater-cones appear under a high sun as minute white spots and can be studied to advantage only with powerful instruments. The Italian astronomer, Maggini, observing the floor of the lunar crater, Plato, in 1916 noted that one of the small crater-cones that exist there in great numbers, was temporarily obscured from view by a cloud of reddish vapors, and Prof. W. H. Pickering, at Arequipa, Peru, observing the same region some years ago, believed that he saw evidence of change in some of these small markings. The crater, Plato, has probably been more carefully studied than any other portion of the lunar surface. It is sixty miles in diameter and may be seen even without a telescope as a dark "eye" not far from the northern edge of the moon. Its floor is one of the darkest objects in the moon—a dark steel-grey in color—and there is no doubt that for some unknown reason its dark hue deepens from the time the sun has an altitude of twenty degrees until after full moon. It has a brilliant white wall rising from 3,000 to 4,000 feet above its floor, crowned with several lofty peaks and intersected by a number of valleys and passes. The spots and faint light markings on the floor have been the object of much study with small or moderate sized instruments, and at least six of them are known to be crater-cones. Since they can only be studied to advantage with powerful instruments and as such instruments are rarely used for a systematic study of lunar markings, it is difficult to settle the controversy as to whether they have changed in appearance or have been at any time obscured by vapors. Most lunar observing is done—necessarily—with smaller instruments because the majority of astronomers appear to have accepted the view that the moon is a dead world, and those who are engaged in astronomical work with our greatest telescopes seem to feel that other fields of research will prove more fruitful. Possibly it is for this reason that we know so little about our nearest neighbor in space! There are at least as many unsolved problems confronting us on the moon as there are among the distant stars.

Geologists tell us that more oxygen is to be found in the first six feet of the earth's crust than in all of the atmosphere above. Does oxygen not exist in the surface rocks of the moon as well?

Volcanic action, we are told, is primarily an escape of gases from the interior, chiefly hydrogen, nitrogen, hydrocarbons, sulphur, and various compounds, as well as vast quantities of steam. Beneath the surface chemical change is continually taking place which results in the release of an enormous amount of heat. Some of the gases mentioned above combine with the oxygen in the surface rocks and heat is evolved. It is a known fact that there is great inherent heat in the earth's surface crust. Why not in the moon's surface crust as well?

The water that would be expelled in the form of steam from volcanic vents on the moon would be transformed immediately into hoar-frost, snow and ice and would settle down upon the flanks of the crater-cones or vents.

It should be borne in mind that only volcanic activity on an enormous scale would be plainly visible to us even with the powerful telescopes at our command. Ordinary eruptions such as occur on our own planet would be very difficult to detect. Since the escaping vapors would rapidly pass into the solid state and settle down upon the flanks of the crater-cones or vents, we would observe in general little if any change in an object unless we chanced to be looking at it at the time of the eruption, when it might appear to be temporarily obscured by a veil of vapors. What are the chances that we would be carefully observing at the precise time of an eruption, a minute marking, two or three miles in diameter, on a surface as large as all of North America, a surface that is covered with some 30,000 charted craters, numberless crater-pits, streaks, rays, spots, clefts and rills in intricate systems, mountain chains and valleys and a mass of intricate detail?

If we were looking at the earth from the moon with the aid of a powerful telescope would we be apt to notice an eruption of Vesuvius or Katmai or Mauna Loa? Objects four or five miles in diameter would appear as hazy spots with nothing distinctive or remarkable in their appearance. Yet vapor and steam arising from terrestrial volcanoes would be carried by our atmosphere over an area of many square miles, while there is no atmosphere on the moon to spread the vapors that may arise from similar volcanic vents. It would have to be a cataclysmic change indeed to be accepted as indisputable evidence that change is taking place on the moon, and the days of gigantic upheavals are probably over on our satellite as well as on the earth. If volcanic activity is still taking place on the moon it is probably in a mild form such as a comparatively quiet emission of gases from volcanic vents and fumaroles. Such forms of activity would not be plainly visible at this distance, even with the aid of powerful telescopes. The problem of detecting changes on the moon is complicated by the fact that a change of illumination greatly alters the appearance of all lunar markings. Such a change is continually taking place in the course of the month. A marking that stands out in bold relief at lunar sunrise or sunset will change entirely in appearance a few days later under a high sun or even disappear from view entirely. These changes in phase or illumination have to be taken account of in the search for evidence of actual change. To decide whether or not change has actually taken place the object must be viewed under similar conditions, so far as they can be obtained. Even when special care is taken in this respect the suspected evidence of change is usually "explained away" as due to differences in illumination or seeing, by those who have not observed the object themselves and are not in sympathy with the view that the moon is anything but a dead world.

As regards the question of life on the moon, it is interesting to consider the facts brought out by investigations made by scientists connected with the Geophysical Laboratory of the Carnegie Institute in the Valley of Ten Thousand Smokes. The volcanic activity there takes the form of eruptions from numerous small vents or fumaroles and ninety-nine per cent. of the emanations are water vapor. It was observed that blue-green algae were living at the edge of active vents emitting ammonia compounds at a temperature of 212° F. They were not found, however, near vents from which ammonia compounds were not being emitted. If life exists under such conditions it is conceivable that suitable conditions for the support of certain forms of life, animal as well as vegetable, may be found in low-lying valleys and crevices and upon the floors of craters, where certain gases essential to the support of life might be evolved from many small volcanic vents and fumaroles.

Many theories have been advanced to explain the origin of the lunar craters which have no counterpart on our own planet. They are saucer-like depressions in the surface of the moon, frequently of such great size that an observer standing in the center would not be able to see either side of the crater owing to the curvature of the moon's surface. Craters fifty, sixty or one hundred miles in diameter are by no means uncommon, while there are thousands between five and fifty miles in diameter. A characteristic feature of many craters is a central peak, and the surrounding walls are often a mile or more high and in some instances are symmetrically terraced. New craters have been formed on the sides or floors of old craters, and these are always more clear-cut and sharper in outlines than the old formations, and generally much smaller. A number of craters are surrounded by a system of light streaks or rays of unknown origin that extend in some instances to enormous distances on all sides of the crater. The most conspicuous system is the one surrounding the lunar crater Tycho near the south pole of the moon. The rays originating in this crater extend in all directions for hundreds of miles without turning aside for any obstructions, passing over mountains, craters and plains in their course in practically straight lines like spokes in a wheel. This ray system of Tycho is the most noticeable marking on the moon's surface at the time of full moon. As these streaks cast no shadow they are apparently cracks in the surface that have filled up with some light-colored material from below. Their origin has never been satisfactorily explained.

As to the origin of the lunar craters, some believe that they were produced in past ages by a bombardment of the lunar surface by huge meteoric masses; but there are many objections to this theory that we will not take up here. It is more generally believed that the lunar craters are a result of volcanic activity on an enormous scale which took place on the moon many ages ago and which has now practically ceased, its only manifestations now taking the form of a quiet emission of gases from small volcanic vents or fumaroles which exist all over the lunar surface but which are to be found in greatest numbers on the floors and sides of craters.

## CHAPTER XXII. COMETS.

The orbits of comets are inclined at all angles to each other and to the orbits of the planets which, on the other hand, lie very nearly in the same plane.

The larger members of the sun's family, the planets and their satellites, revolve from west to east around the sun. Comets on the contrary frequently retrograde or back around the sun in the opposite direction—from east to west.

The paths that these erratic visitors follow in their journeys around the sun bear not the slightest resemblance to the paths of the planets, which are almost perfect circles. The orbits of comets are ellipses that are greatly elongated or parabolas. If the orbit is a parabola the comet makes one and only one visit to the sun, coming from interstellar space and returning thereto after a brief sojourn within our solar system.

Donati's comet of 1858, one of the greatest comets of the nineteenth century, had a period of more than two thousand years and its aphelion (the point in its orbit farthest away from the sun) was five times more distant than the orbit of Neptune.

There is, however, a class of comets known as periodic comets that have extremely short periods of revolution around the sun. To this class belongs Halley's comet whose period of seventy-five years exceeds that of any other short period comet. Encke's comet, on the other hand, has a period of three and a third years which is the shortest cometary period known. Most of the periodic comets are inconspicuous and only visible telescopically even when comparatively near to the earth. Halley's comet is the only one of this class that lays any pretensions to remarkable size or brilliancy and it also is showing the effects of disintegration resulting from too frequent visits to the sun.

Comets are bodies of great bulk or volume and small total mass. Their tails, which only develop in the vicinity of the sun, are formed of the rarest gases, and the best vacuum that man can produce would not be in as tenuous a state as the material existing in the tails of comets. There are many proofs of the extreme tenuity of comets. The earth has on a number of occasions passed directly through the tails of comets without experiencing the slightest visible effects. Stars shine undimmed in luster even through the heads of comets. If the earth should encounter a comet "head on" it is doubtful if it would experience anything more serious than a shower of meteors which would be consumed by friction with the earth's atmosphere, or a fall of meteorites over a small area of a few square miles. It is possible, however, that matter in the nucleus, the star-like condensation in the head of a comet, may consist of individual particles weighing in some instances a number of tons, surrounded by a gaseous envelope and held together by the loose bonds of their mutual attraction. If the earth should encounter the nucleus of a comet considerable damage might be done over a portion of the earth's surface, but the chances of such an occurrence are less than one in a million.

Since the total mass of a comet is so small, a close approach to one of the planets, especially Jupiter, produces great changes in the form of the comet's orbit, though the motion of the planet is not disturbed in the slightest degree by the encounter.

The majority of all the short-period comets have been "captured" by Jupiter, that is, the original orbits have been so changed by the perturbations produced by close approaches to the giant planet that their aphelia, or the points in their orbits farthest from the sun, lie in the vicinity of Jupiter's orbit. Several of the other planets have also "captured" comets in this sense, and the fact that the aphelia of a number of comets are grouped at certain definite intervals beyond the orbit of Neptune has been considered by some astronomers to be an indication that there are two or more additional planets in the solar system revolving around the sun at these distances.

The most interesting feature of a comet is its characteristic tail which develops and increases in size and brilliancy as the comet approaches the sun. As the tail is always turned away from the sun it follows the comet as it draws near the sun and precedes it as it departs. Its origin is due, it is believed, both to electrical repulsion and light-pressure acting upon minute particles of matter in the coma or head of the comet.

The curvature of the tail depends upon the nature of the gases of which it is composed. Long, straight tails consist chiefly of hydrogen, it has been found, curved tails of hydrocarbons and short, bushy tails of mixtures of iron, sodium and other metallic vapors. At times the same comet will have two or more tails of different types.

Since the material driven off from the nucleus or head of a comet by electrical repulsion and light-pressure is never recovered, it is evident that comets are continually disintegrating. Also, comets that have passed close to the sun at perihelion have frequently been so disrupted by tidal forces that one nucleus has separated into several parts and the newly formed nuclei have pursued paths parallel to the original orbit, each nucleus developing a tail of its own.

Many periodic comets, it is now known, have gradually been broken up and dissipated into periodic swarms of meteors as a result of the disruptive effect produced by too frequent returns to the vicinity of the sun.

These swarms of meteors continue to travel around the sun in the orbits of the former comets. The earth encounters a number of such swarms every year at certain definite times.

The largest and best known of these swarms or showers are the Leonids, which appear about November 15; the Andromedas (or Bielids), which appear later in the same month and the Perseids, which appear early in August. These swarms are named for the constellations in which their "radiant" lies, that is, the point in the heavens from which they appear to radiate. The position of the radiant depends upon the direction from which the swarm is coming. It is simply a matter of perspective that the individual particles appear to radiate from the one point, for they are actually travelling in parallel lines.

The luminosity of these meteoric particles is caused by the friction produced by their passage through the atmosphere. They always appear noiselessly because they are mere particles of meteoric dust weighing at the most scarcely a grain. They differ greatly in this respect from their large and noisy relatives, the meteorites, bolides and fireballs.

Numberless small meteoric particles are entrapped by the earth's atmosphere every day. They are referred to as "shooting" stars or "falling" stars though, of course, they are not in any sense stars. It is only when these meteoric particles travel in well-defined cometary orbits and appear at certain definite times every year that they are referred to as swarms or showers of meteors.

The luminosity of comets is due not only to reflected sunlight, but to certain unknown causes that produce sudden and erratic increases or decreases of brilliancy. The causes of these sudden changes in luminosity are unknown; possibly electrical discharges or chance collisions between fragments of considerable size may account for some of them.

The peculiar behavior of the tails of comets at certain times has frequently been noted and suggests the existence of quantities of finely-divided meteoric or gaseous matter within the solar system that has no appreciable effect upon the huge planetary masses, but offers sensible resistance to the passage of the tenuous gases of which the tails of comets are composed. The fact that the earth daily encounters meteoric dust, meteorites and fireballs also indicates that meteoric matter exists in considerable quantities within our solar system. Tails of comets appear at times to be twisted or brushed aside as if they had encountered some unknown force or some resisting medium.

Up to the present time several hundred comets have been discovered. Nearly three-fourths of this number travel in orbits that appear to be parabolas. Of the remaining number there are about forty that have been "captured" by the major planets, Jupiter, Saturn, Uranus and Neptune, though Jupiter possesses the lion's share of these captured comets. Scarcely a year passes by that several comets are not discovered. Most of these are telescopic, however, even when they are near the sun and at their greatest brilliancy. Naked-eye comets of great splendor and brilliancy are comparatively rare and there has been a particular dearth of such unusual comets during the past thirty years or so.

The last spectacular comet, unless we make an exception of Halley's periodic comet, which made its return according to prediction in 1910, was the great comet of 1882 which was visible in broad daylight close to the sun and at its perihelion passage swept through the solar corona with a velocity that exceeded two hundred and fifty miles a second and carried it through one hundred and eighty degrees of its orbit in less than three hours.

Some comets approach much closer to the sun than others. The majority of all comets observed have come within the earth's orbit and no known comet has its perihelion beyond the orbit of Jupiter. It is, of course, possible that there may be a number of comets that never come within the orbit of Jupiter, but it is very unlikely that any such comet would ever be discovered. The majority of comets are simply small, fuzzy points of light that are only visible telescopically and the greater the perihelion distance of the comet the less likely is it to be seen with the naked eye.

Since comets as well as planets obey Kepler's first law, known as the law of areas, and sweep over equal areas in equal times, it is evident that when a comet is at perihelion, or nearest to the sun, it is moving at maximum speed and when it is at aphelion, or farthest from the sun, it is moving at minimum speed. Moreover, its speed at these two points in its orbit varies tremendously since the orbits of comets are ellipses of very high eccentricity. The speed with which the planets are traveling is, on the other hand, remarkably uniform since their orbits are nearly circular.

The leisurely speed with which a comet travels through the frigid outer regions of the solar system is gradually accelerated as the comet draws nearer and nearer to the sun until it has acquired near the time of perihelion passage a velocity that occasionally exceeds two hundred miles a second. Here, also, the great increase in light and heat and the strong magnetic field of the sun produce complex changes in the gaseous and meteoric substances of which the comet is composed until the characteristic tail and peculiar cometary features are fully developed. As the comet again recedes from the sun after perihelion passage its speed slackens once more. It soon parts with its tail and other spectacular features and fades rapidly from view even in the largest telescopes.

## CHAPTER XXIII. METEORITES.

Meteorites, bolides or fireballs, as they are variously called, are stones that fall to the earth from the heavens. They furnish the one tangible evidence that we possess, aside from that furnished by the spectroscope, as to the composition of other bodies in space and it is a significant fact that no unknown elements have ever been found in meteorites, though the forms in which they appear are so characteristic that they make these stones readily distinguishable from stones of terrestrial origin.

The origin of meteorites is not definitely known, but the evidence is very strong in favor of the theory that they are the larger fragments of disintegrated comets of which meteors and shooting stars are the smaller; the distinction between the two being simply that the latter class includes all bodies that are completely consumed by friction with the earth's atmosphere and, therefore, only reach the surface in the form of meteoric dust.

According to other theories meteorites may be fragments of shattered worlds that have chanced to come too near to a larger body and have been disrupted, or they may possibly be the larger fragments of the disintegrated comets of which the meteoric swarms are the smaller.

Interplanetary space is not altogether a void. Our own planet sweeps up in the course of a single day, it has been estimated, approximately twenty million shooting stars or meteors of sufficient size to be visible to the naked eye, while the estimate for the telescopic particles runs up to four hundred million.

Meteorites on the other hand are comparatively rare. On the average it has been estimated about one hundred meteorites strike the earth in the course of a year, of which number only two or three are actually seen. According to Bulletin 94, U. S. National Museum, approximately six hundred and fifty falls and finds of meteorites have been reported, representatives of which appear in museums and private collections.

Meteorites, as well as shooting stars and meteors, frequently appear in showers. In such instances the fall usually consists of several hundred or thousand individual stones and the area over which they fall is several square miles in extent and roughly ellipsoidal in shape. One of the most remarkable of such falls  hundred thousand stones, varying in weight from fifteen pounds to a small fraction of an ounce, fell near Pultusk, Poland. Another remarkable fall of meteorites occurred at

L'Aigle, France, in 1803. Between two thousand and three thousand stones fell over an ellipsoidal area of six and two-tenths miles in greatest diameter, the aggregate weight of the stones being not less than seventy-five pounds.

This fall of stones is of particular interest since it took place at a time when men were still very doubtful as to whether or not stones actually fell to earth from the heavens.

After this fall had occurred in a most populous district of France in broad daylight and attended by violent explosions that lasted for five or six minutes and were heard for a distance of seventy-five miles, no reasonable doubt could longer be held as to the actuality of such phenomena.

Meteorites are without exception of an igneous nature, that is, they are rocks that have solidified from a molten condition. They can be classified into three groups, Aerolites or Stony Meteorites, Siderolites or Stony-iron Meteorites, and Siderites or Iron Meteorites.

More iron meteorites seem to have fallen in Mexico and Greenland than in any other part of the world—at least of its land surface.

Yet strange to say, of all the meteorites that have been seen to fall only nine belong to the group of Siderites or Iron Meteorites, though the three largest meteorites known, Peary's meteorite from Cape York, Greenland, weighing 37½ tons, the meteorite lying on the plain near Bacubirito, Mexico, weighing about 20 tons, and the Willamette, Oregon, meteorite, weighing 15½ tons all belong to this group. Moreover, all the Canyon Diablo meteorites, which are strewn concentrically around Coon Mountain crater in northern Arizona to a distance of about five miles, are members of this same group. Coon Mountain or Meteor crater itself is a perfectly round hole, about six hundred feet deep and over four thousand feet in diameter and was formed, it is believed, by the impact of a huge meteorite which has never been found. It is believed that the Canyon Diablo meteorites, of which there are nearly four hundred individuals in the U. S. National Museum alone, were all members of this same fall. It is possible that these meteorites of the Canyon Diablo district, with the huge meteorite that produced the crater itself, formed the nucleus of a comet that struck the earth not more than five thousand years ago, according to the geological evidence.

All iron meteorites or siderites (from the Greek sideros, iron) are composed chiefly of alloys of nickel and iron. The percentage of nickel in these iron meteorites is very small, usually from five to ten per cent., while the iron forms about ninety or ninety-five per cent. of the whole. Cobalt is also present in practically all iron meteorites in small quantities of 1 per cent. or less. Usually small quantities of iron sulphide and phosphide as well as graphite or some other form of carbon appear in the iron meteorites and in some instances black and white diamonds have been found, as in some of the Canyon Diablo irons.

A very interesting and beautiful feature of many iron meteorites is the Widmanstätten figures which appear when a section of such a stone is polished and treated by means of a weak acid. These figures are due to the unequal solubility of the three different alloys of nickel and iron of which the stones are composed. The irons giving the Widmanstätten figures are known as octahedral irons. Other irons known as hexahedral irons give figures of a different type known as Neumann figures when the polished section is treated with weak acid, while other irons are so homogeneous in their composition that they show no figures at all.

Aerolites or Stony Meteorites occur more abundantly than iron or stony-iron types, and they are classified into many divisions and subdivisions according to their composition. In these stones appear certain compounds that are commonly met with in terrestrial igneous rocks. The mineral that is most abundant in the stony meteorites, composing sometimes nearly seventy-five per cent. of the stone, is a magnesium and iron silicate known as olivine, which is also usually present in terrestrial rocks of an igneous nature. Certain compounds found in the stony meteorites are rarely if ever found in terrestrial rocks, however, and these serve to distinguish the stony meteorites readily from stones of terrestrial origin. The alloys of iron and nickel, for instance, that occur in minor quantities in the stony meteorites and make up usually about ninety-five per cent. of the mass of the iron meteorites, are never found in terrestrial rocks. Although about thirty of the terrestrial elements are to be found in meteorites, the forms and compounds in which they appear are so characteristic and on the whole so different from those occurring in terrestrial rocks, that the analyst has no difficulty in distinguishing between the two. There are, for instance certain formations known as chondrules, peculiar spherical and oval shapes, varying in size from minute particles to objects the size of walnuts, appearing in many varieties of stony meteorites that are never found in terrestrial rocks, and that are one of the most puzzling features associated with the origin and nature of these stones. Sometimes the chondrules are so loosely embedded in the stone that they fall away when it is broken. In some instances almost the entire stone is made up of these chondrules. According to one theory the chondrules were originally molten drops, like fiery rain, and their internal structure, which is greatly varied, depends upon their conditions of cooling.

Stony meteorites, in which these chondrules are to be found, are spoken of as chondrites. There are white and gray and black chondrites and crystalline and carbonaceous chondrites, according to the nature of the chondrules found in the stones.

Stony meteorites also contain minute quantities of iron and nickel alloys in the form of drops or stringers.

Upon entering the earth's atmosphere stony meteorites become coated with a thin black crust which is a glass formed by the fusion of its surface materials by the heat generated during its passage through the atmosphere.

In many of the stony meteorites there also appear fine thread-like veins which are due to the fracturing of the stone prior to its entrance into the atmosphere. The material filling these veins is coal black in color, opaque and of an unknown composition.

Many meteorites show signs of collisions and encounters with other meteorites outside of the atmosphere as would be expected as they travel in swarms and groups. Sometimes the entire meteorite is composed of fragments of two or more distinct stones cemented together. Such a stone is spoken of as a breccia.

In the third class of meteorites to which we now come, known as the stony-iron meteorites, there is a network or sponge of nickel-iron alloy, the interstices of which are filled with stony material.

When this network or sponge is continuous the meteorite is spoken of as a stony-iron pallasite. When the network of metal is more or less disconnected the meteorite is a meso-siderite.

If meteorites are heated in a vacuum, the conditions existing in interplanetary space being thus produced to a certain extent, they give forth their occluded gases and it has been found that these gases give spectra identical with the spectra of certain comets. Meteoric irons give forth hydrogen as their characteristic gas while the gases occluded in the stony meteorites are chiefly the oxides of carbon, carbon monoxide and carbon dioxide. It has been found that the amount of gases contained in a large meteorite or shower of meteorites is sufficient to form the tail of a comet. These facts all tend to strengthen the belief that meteorites are indeed cometary fragments.

In view of the fact that some geologists believe meteorites may be fragments of other worlds, it is of interest to know that so far no fossil-bearing meteorites have been found, and if meteorites are fragments of a shattered world, such worlds must have been reduced to a molten condition at the time of the catastrophe.

The rapid passage of the meteorite through the air leaves a partial vacuum in its trail into which rush the molecules of air from all sides, producing the characteristic noises that accompany the passage of a meteorite, which have been variously compared to the rattle of artillery, the distant booming of cannons or the rumble of thunder.

There may be, also, explosions of inflammable gases occluded in the crevices of the meteorite which will shatter it into fragments or the meteorite may be shattered by the resistance and pressure of the atmosphere or as a result of the extremes of temperature existing between the interior and its surface. Many meteorites have actually been seen to burst into fragments in the air with a loud report.

There is practically no foundation for the belief that germs of life have been brought to our planet on such igneous rocks. No microscopic examinations of meteorites have yielded any results that could be interpreted in favor of such a view.

Falls of meteorites are accompanied in nearly every instance by terrific explosions and sharp reports that can be heard for many miles around, often causing the ground to shake as in an earthquake. The meteorite itself has been described as resembling a ball of fire or the headlight of a locomotive, and is followed frequently by a trail of light or a cloud of smoke. At the time it enters our atmosphere a meteorite is moving with planetary velocity ranging from two to forty-five miles per second. Its interior is intensely cold, approaching in temperature the absolute zero of interplanetary space, and it is, therefore, far more brittle than it would be at ordinary temperatures. As it ploughs its way into the earth's atmosphere its surface temperature is soon raised by friction to at least 3,000° or 4,000° C., which is sufficient to fuse all surface materials into the characteristic black crust, with which stony meteorites are coated.

Meteorites are usually first seen at an altitude of fifty or sixty miles. Although they are moving with a velocity comparable to that of the planets, when they enter the earth's atmosphere, this velocity is so rapidly reduced by friction with the atmosphere that they usually drop to the surface of the earth with a velocity about equal to that of ordinary falling objects.

The flight of a meteorite often extends over a path several hundred miles in length and the meteorite may be seen by many observers in several different States and yet finally fall in some unknown spot and never be found.

The evidence gathered regarding the actual fall of meteorites is often contradictory. Some stones are too hot to handle for hours after they fall, others are merely warm, while still others have been picked up cool or even intensely cold. Meteorites have been seen to fall upon dried grass and upon straw without producing even charring effects. The evidence regarding the depths to which meteorites penetrate the ground is quite as conflicting. The largest of all the stony meteorites which fell at Krnyahinya, Hungary, weighed 647 pounds and buried itself to a depth of eleven feet. Yet Peary's Cape York iron meteorite, weighing 37½ tons, was only partially covered and showed no signs of abrasions of surface resulting from the fall.

The Willamette iron meteorite, weighing 16½ tons, lay in a forest when found and was not deeply buried. The Bacubirito iron meteorite, weighing 20 tons, lay in soft soil, barely beneath the level of the surface. On the other hand a fragment of a stony-iron meteorite, weighing 437 pounds, that fell at Estherville, Iowa, buried itself eight feet in stiff clay.

Geologists in charge of the meteoric collections of various museums quite frequently have stones sent to them for analysis that are reputed to be of celestial origin. More often than not such stones are found to be purely terrestrial in their origin. The composition of a meteorite is so characteristic and unique that such a stone can never be mistaken. Finds of bona-fide meteorites are on the whole extremely rare.

It is also a peculiar fact that meteorites are usually observed in the months when ordinary meteors or periodic swarms of meteors are least prevalent, that is in the months of May, June and July.

## CHAPTER XXIV. THE EARTH AS A MAGNET.

If a small, freely suspended compass needle is moved over a highly magnetized steel sphere, it will be seen that it constantly changes its position both horizontally and vertically so as to lie always along the "lines of force" of the sphere.

There will be one point on the sphere which we will call the North Magnetic Pole, where the north-seeking end of the needle will point vertically downward or make a "dip" of 90° with the tangent plane. At the diametrically opposite point on the sphere, called the South Magnetic Pole, the opposite end of the compass, the south-seeking end, will point vertically downward; while at a point midway between the magnetic poles of the sphere the needle will lie parallel to the diameter connecting the two poles and there will be no dip.

The total intensity of the magnetic field surrounding the sphere will be found to be greatest in the vicinity of the magnetic poles and least, midway between the poles.

Now, a freely suspended compass needle carried to all parts of the earth will behave very much in the same manner as the needle moved over the magnetized steel sphere. There are two points on the earth's surface, known as the North and South Magnetic Poles, where the needle points vertically downward and approximately midway between is the Magnetic Equator where the compass needle places itself in a perfectly horizontal position and the "dip" of the needle is zero. In other words, the earth acts as a huge magnet and possesses a magnetic field with lines of force converging towards its poles similar to the lines of force of the steel sphere.

There are, however, some very important differences between the sphere of steel and our earth. The matter of which the earth is composed is not homogeneous. It is believed to possess an iron core of considerable size, it is true, but its outer shell is composed of heterogeneous masses that in certain regions cause very appreciable local deflections of the needle. It is surrounded, moreover, by an atmosphere permeated by electrified particles of matter shot forth from the sun, which we now know is a still greater magnet surrounded by a magnetic field that is of the order of 50 gausses at the poles and about eighty times more powerful than that of the earth.

It is now a well-established fact that the sun's magnetic field exerts a powerful influence over the condition of the earth's magnetic field, and that vast solar disturbances affect very materially the direction and intensity of the lines of force.

It is thus little wonder that this non-homogeneous and rapidly rotating terrestrial globe, surrounded by an electrified atmosphere and subject to the action of a still more powerful magnet, the sun, should not behave in a manner exactly analogous to a uniformly magnetized steel sphere.

The earth's magnetic poles are neither symmetrically placed nor absolutely fixed in position. There is every reason to suspect that they shift about from year to year, and possibly fluctuate irregularly in position in the course of a few days or hours under the influence of disturbing forces. The position of the earth's North Magnetic Pole, last visited by Amundsen in 1903, now lies approximately in Latitude 70° N. Longitude 97° W. The position of the South Magnetic Pole, according to the latest determinations, is, in round numbers, in Latitude 73° S. and Longitude 156° E. of Greenwich. It is evident, therefore, that the magnetic poles of the earth are not symmetrically placed and that they lie fully 30° from the geographical poles. The chord connecting the magnetic poles passes 750 miles from the earth's center, and it is about 1,200 miles from the geographic pole to the nearest magnetic pole. There exist, moreover, in high latitudes local magnetic poles, due possibly to heavy local deposits of ore. One such pole was discovered at Cape Treadwell, near Juneau, Alaska, during Dr. L. A. Bauer's observations there in 1900 and 1907. In the center of the observing tent at this point the needle pointed vertically downward and the compass reversed its direction when carried from one side of the tent to the other.

It is a well-known fact that there are very few points on the earth's surface where the compass needle points either to the true geographical pole or to the magnetic pole, and if it does chance to do so, it is a transient happening. The "variation of the compass" or the declination of the needle, as it is called, is the angle that the compass needle makes with the true north and south line or the meridian. It is an angle of greatest importance to navigators and explorers, for it gives them their bearings, yet it is unfortunately subject to ceaseless variations of a most complicated nature, since it depends on the constantly pulsating and never ceasing magnetic changes that sweep over the surface of the earth and through its crust. It is affected by long period or secular changes, as they are called, progressing more or less regularly in obscure cycles of unknown period. It is subject to a diurnal change that depends on the position of the sun relative to the meridian, and that varies with the seasons and with the hour of the day. It is affected by the sun spot cycle of 11.3 years which has a direct effect upon the intensity of the earth's magnetic field. The intensity of the magnetic field in sun spots is, according to Abbot, sometimes as high as 4,500 gausses or 9,000 times the intensity of the earth's field. At times of maximum spottedness of the sun the intensity of the earth's magnetic field is reduced.

Moreover, when great and rapidly changing spots appear upon the sun, electrified particles are shot forth from the sun with great velocity and in great numbers, and are drawn in towards the magnetic poles of the earth. Meeting the rarefied gases of the earth's upper atmosphere, they illuminate them as electric discharges illuminate a vacuum tube. Some of these electrons are absorbed by gases at high elevations, other descend to lower levels. The most penetrating rays have been known to descend to an altitude of twenty-five miles which is about the lowest limit yet found for auroral displays. It is the passage of these rays through the atmosphere that cause the magnetic disturbances known as magnetic storms, that are associated with the appearance of great sun spots and auroral displays. At such times sudden changes take place in the intensity of the earth's magnetic field that cause the compass needle to shiver and tremble and temporarily lose its directive value. These magnetic storms have been known to produce great temporal changes in the intensity of the earth's field. According to Dr. L. A. Bauer, Director of the Department of Terrestrial Magnetism of the Carnegie Institute of Washington, the earth's intensity of magnetization was altered by about one-twentieth or one-thirtieth part by the magnetic storm of September 25, 1909, which was one of the most remarkable on record, and the earth's magnetic condition was below par for fully three months afterwards as a result.

In addition to these various regular and irregular changes in the variation of the compass, or declination of the needle, due to changes in the earth's magnetic field as a whole, there are local effects due to restricted regional disturbances of the earth's magnetic field or to local deposits of ore, or to volcanoes or other local causes. The effect of all these disturbances upon the declination of the needle must be determined by continual magnetic surveys of all portions of the earth's surface.

As a whole the earth's magnetic field is more uniform over the oceans than over the land, with all its disturbing topographical features. Yet this advantage is offset largely in navigation by the fact that every steel ship that sails the seas is a magnet, with its two magnetic poles and its neutral line where the two opposite magnetic forces are neutralized, as is the case with every magnet. The direction in which a steel ship lies with reference to the earth's magnetic field while it is being built determines the position of the magnetic poles in its hull and the position of its neutral line and this distribution of magnetism over a ship's hull must be taken account of in the installation of its standard compass. Every piece of horizontal and vertical iron aboard ship has an effect upon the variation of the compass and compensation must be made for such disturbing forces. The direction of sailing, the position in which a ship lies at dock, storms encountered at sea, the firing of batteries (on warships) are some of the factors that are operative in producing changes in the variation of the magnetic compass aboard a ship.

Every ship must undergo at frequent intervals magnetic surveys for the purpose of determining its magnetic constants and its "Table of Deviations of the Compass."

The direction in which the compass needle points aboard ship is the resultant of the effect of the earth's magnetic field and the magnetic field of the ship, and both fields are subject to continual and complicated variations from year to year, from day to day, and even from hour to hour!

The elements of the earth's magnetic field are determined for any one epoch by long-continued magnetic surveys carried on to a greater or less extent by the various nations of the world, and the results are published in the form of magnetic charts for land and sea, showing the values of the three magnetic elements, declination of the needle, dip or inclination, and horizontal intensity of the earth's field for a definite period. So rapid are even the long-period changes in the earth's magnetic field that a magnetic chart can be relied upon for only a very few years and fresh data for the construction of these charts that are so valuable to navigators and explorers must be gathered continually.

The Department of Terrestrial Magnetism of the Carnegie Institute of Washington is engaged in continual magnetic surveys of the earth by land and sea that are of the highest value not only to navigators but also to scientists interested in solving the great and mysterious problem of the underlying causes of the earth's magnetism.

To give an idea of the extent and scope of the work of this department it may be mentioned that its non-magnetic ship Carnegie made in the period 1909-1918 a total run of 189,176 nautical miles, nearly nine times the earth's circumference, with an average day's run of 119 nautical miles. Magnetic observations were made practically every day at a distance of 100 to 150 miles apart. In this nine-year period five cruises were made. On her first cruise the Carnegie sailed from St. John's, Newfoundland, to Falmouth, England, over practically the same course followed by the famous astronomer, Halley, in the Paramour Pink two centuries earlier to determine the variation of the compass. In her fourth voyage the Carnegie circumnavigated the world in sub-antarctic regions in 118 days—a record time. She has traversed all oceans from 80° North to the parallel of 60° South and has crossed and recrossed her own path and the path of her predecessor, the Galilee, many times, thus making it possible to determine for the points of intersection the secular changes in the magnetic elements.

After a thorough overhauling in 1919 and the installation of a four-cylinder gasoline engine, made of bronze throughout, to take the place of the producer-gas engine used on earlier cruises, the Carnegie started on her sixth cruise with a crew of twenty-three officers and men on October 9, 1919. A cruise of 61,500 miles was planned in the South Atlantic, Indian and Pacific Oceans to last approximately two years. Unsurveyed regions in the South Atlantic and Indian Ocean were to be covered and the route was planned so as to obtain a large number of observations of the progressive changes that have taken place in the magnetic elements. This is accomplished as stated above by intersecting former routes and obtaining new values of the element at the points of intersection.

In addition to its ocean magnetic surveys the Department of Terrestrial Magnetism also carries on extensive land surveys in all parts of the globe. In 1919 special expeditions were sent out by the Department to observe the total solar eclipse of May 29th at stations distributed over the entire zone of visibility of the eclipse and immediately outside. At Dr. Bauer's station in Liberia the total phase was visible in a cloudless sky for more than six minutes, which is very close to the maximum length of phase that can possibly be observed. Unmistakable evidence was gathered at all stations of an appreciable variation in the earth's magnetic field during a solar eclipse, which variation is the reverse of that causing the daylight portion of the solar diurnal variation of the needle.

In addition to the magnetic survey work on land and sea which is the chief work of the Department of Terrestrial Magnetism, atmospheric-electric observations are carried on continually on land and sea and experiments have been carried on at Langley Field, Va., lately, in the development of methods and instruments for determining the geographical position of airplanes by astronomical observations. There has also been recently formed under this department a Section of Terrestrial Electricity.

The cause of the earth's magnetic field is still one of the greatest unsolved problems of astro-physics. The theory that has been advanced by Schuster that all large rotating masses are magnets as a result of their rotation has received considerable attention from astrophysicists, and attempts have been made to prove this experimentally. It has been found that iron globes spun at high velocities in the laboratory do not exhibit magnetic properties. This may mean simply that the magnetic field is too weak to be detected in the case of a comparatively small iron sphere spun for a limited period under laboratory conditions. It must be remembered that the earth has been rotating rapidly on its axis for millions of years and is, compared to terrestrial objects, an extremely large mass. Yet it has been shown that as a whole our earth is an extremely weak magnet, and that if it were made entirely of steel and magnetized as highly as an ordinary steel-bar magnet, the magnetic forces at its surface would be a thousand times greater than they actually are.

If it is true that all rotating bodies are magnets, then all the heavenly bodies, planets, suns and nebulæ are surrounded by magnetic fields. We know nothing to the contrary. In fact, we know this to be true for the earth and sun, and strongly suspect that it is so in the case of the planets Jupiter and Saturn.

When we understand more about the properties of matter, the nature of magnetism, as well as of gravity, may be revealed to us.

## CHAPTER XXV. SOME EFFECTS OF THE EARTH'S ATMOSPHERE UPON SUNLIGHT.

It is impossible to exaggerate the importance of the atmosphere to all forms of life upon the surface of the earth. If there were no atmosphere there would be no life, because it is through the agency of the water-vapor, carbon-dioxide and oxygen in the atmosphere that all life-processes are maintained.

If there were no atmosphere there would not only be no life upon the earth; there would be also none of the beautiful color effects produced by the passage of sunlight through the atmosphere. There would be no blue skies, no beautiful sunrise and sunset effects, no twilight, no rainbows, no halos, no auroral displays, no clouds, no rains, no rivers nor seas, no winds nor storms. The heavens would be perfectly black except in the direction of the heavenly bodies which would shine as brilliantly by day as by night.

To understand how the atmosphere produces color effects such as blue skies, sunrise and sunset tints, rainbows and halos, as well as the twinkling of the stars, and numerous other phenomena, we must know something of the nature of light itself.

Light moves outward from any source, such as the sun, in all directions radially, or along straight lines (so long as it does not encounter a gravitational field) with the unimaginable velocity of 186,000 miles per second. As it advances it vibrates or oscillates back and forth across its path in all directions at right angles to this path, unless it is plane polarized light, in which case its vibrations are confined to one plane only.

These vibrations or oscillations of light take the form of a wavelike motion, one wave-length being the distance passed over in the time of one vibration, measured from crest to crest or from trough to trough of adjacent waves.

We may consider that a beam or ray of sunlight is made up of a great number of individual rays of different wave-lengths and different colors. The average wave-length of light, the wave-length of the green ray in sunlight, is about one-fifty-thousandth part of an inch, that is, it would take about fifty-thousand wave-lengths of green light to cover a space of one inch. Now, since light makes one vibration in passing over a distance of one wave-length, it makes fifty thousand vibrations, while advancing one inch, and since it advances one hundred and eighty-six thousand miles in one second we can easily figure out that a ray of sunlight of average wave-length makes about six hundred trillion vibrations (600,000,000,000,000) in a single second!

The chief colors of which sunlight or white light is composed are red, orange, yellow, green, blue, indigo and violet, though there are an infinite number of gradations of color which blend into one another, gradually producing the intermediate tints and shades. The colors just mentioned are called the primary colors of the solar spectrum, which can be produced as a band of light of variegated colors, arranged in the order named by passing a ray of ordinary sunlight through a glass prism. The individual rays of different color and wave-length that make up a beam of sunlight, or white light, then separate out in the order of the wave-lengths. The red rays vibrate the most slowly and have the longest wave-length of all the rays of the visible spectrum. About four hundred trillion vibrations of red light reach the eye in one second. Violet rays, on the other hand, vibrate the most rapidly of all the visible rays and have the shortest wave-length. About eight hundred trillion vibrations of violet light reach the eye every second. The wave-lengths of the intermediate colors decrease in length progressively from the red to the violet and, of course, the frequencies of their vibrations increase in the same order. All sunlight is made up of these rays of different colors and different vibration frequencies, and of other rays as well, to which the human eye is not sensitive, and which, therefore, do not appear in the visible spectrum. Among these invisible rays are the infra-red rays which come just below the red of the visible spectrum, and which are of longer wave-length than the red rays, and the ultra-violet rays, which lie beyond the violet rays of the visible spectrum, and are of shorter wave-length than the violet rays.

Now a ray of ordinary sunlight is separated into the rays of various colors, which form the solar spectrum when it passes from a medium of one density obliquely to a medium of another density, as when it passes from air to glass, or from air to water, or from outer space into the earth's atmosphere. Under such circumstances its velocity is slowed down when it passes from a rare to a denser medium, and the waves of different wave-lengths are bent from their former course, or refracted, by different amounts. The red rays, of longest wave-length, are bent from their former course the least, and the violet rays, of shortest wave-length, are bent the most upon passing from a rare to a denser medium. As a result the ray of sunlight is spread out or dispersed into its rays of different wave-length and color upon entering a medium of different density. It is this refraction and dispersion of sunlight that produces many color effects in the earth's atmosphere.

The atmosphere is not of uniform density throughout. At high altitudes it is extremely rare. That is, there is little of it in a given volume. Close to the earth's surface, however, it is comparatively dense. Half of all the atmosphere is within three and one-half miles of the surface and half of the remainder lies within the next three and a half miles. We may consider it as made up, on the whole, of layers of different densities, strongly compressed near the surface.

Imagine a ray of sunlight entering the earth's atmosphere from without. If it comes from a point in the zenith its course is not changed upon entering the atmosphere, because light passing from a certain medium—as space—into a medium of different density, is not bent from its course, or refracted, provided it enters the new medium in a direction perpendicular to the surface. If it enters the atmosphere (which is the new medium of greater density) obliquely, refraction, or bending of the ray, takes place, and as the ray advances toward the earth, through layers of increasing densities, it is bent from its former course more and more. As the advancing rays of different colors and wave-lengths in the beam of sunlight are slowed down in the new medium, the red rays are turned from their course the least and the violet rays the most and the entire advancing wave-front of the beam of sunlight is bent down more and more toward the horizon, as it proceeds through the atmosphere. As we on the earth's surface see the ray not along its bent course through the atmosphere, but in the direction in which it finally enters our eyes, the effect of refraction upon a ray of light passing through the atmosphere is to displace the object in the direction of the zenith or increase its distance above the horizon. As a result of refraction we see the sun—or moon—above the western horizon after it has really set, and above the eastern horizon before it has really risen. The oval shape that the sun, or moon, often presents on rising or setting, is due to the fact that the light from the lower limb is passing through denser air than the light from the upper limb, and so is refracted more. As a result the lower limb is lifted proportionately more than the upper limb. This distorts the form of the solar or lunar disk, making it appear oval instead of circular.

The familiar twinkling or scintillation of stars and, more rarely, of the planets, is a result of interference of light waves due to irregular and variable refraction in air that is not uniform in density, owing to the presence of constantly rising and descending atmospheric currents of different densities. This also produces the shimmering or unsteadiness of star images in the telescope, that interferes so greatly with accurate measurements of angles or observations of planetary markings.

One may ask why it is, if light from an object, say a star, is bent from its course and separated into rays of various colors upon entering the earth's atmosphere, that we do not see the object drawn out into a band of spectral colors. It is because the angular separation of the various colors is so slight under ordinary circumstances that light from one point is blended with light from a neighboring point of complementary color to produce white light again. Under certain circumstances, however, beautiful color effects may be seen in the earth's atmosphere as a result of the refraction of sunlight.

The blue color of the sky and its brightness is caused by the scattering of the rays of shortest wave-length, the violet and blue rays, by the oxygen and nitrogen in the upper atmosphere. The molecules of these gases interfere with the passage of these rays, powerfully scattering and dispersing them, and thus increasing the length of their path through the air and diffusing their color and brightness in the upper atmosphere, while permitting rays of longer wave-length, the red and orange, to pass on practically undisturbed.

When an object in the heavens lies close to the horizon, the rays of light from it have to travel a longer path through the atmosphere than when the object is overhead, and that too through the densest part of the atmosphere, which lies close to the earth's surface, and in which are floating many dust particles and impurities from the earth's surface. All these particles, as well as the increased density of the atmosphere, interfere with the free passage of the rays, especially of shorter wave-lengths. The violet and blue rays are sifted out and scattered in their long journey through the lower strata of air, far more than when they come to us from an object high in the sky. Even the red and yellow rays are more or less scattered and bent aside—diffracted—by these comparatively large particles near the surface. The reddish color of the sun, moon and even of the stars and planets, when seen near the horizon, as well as the beautiful sunset tints, in which reds and pinks and yellows predominate, are due to the fact that the rays of longer wave-length are more successful in penetrating the dense, dust-laden layers of the lower atmosphere. It is to be free of the dust and impurities as well as the unsteadiness of the lower atmosphere, that observatories are built at high altitudes whenever possible.

When there have been unusually violent volcanic eruptions, and great quantities of finely divided dust have been thrown into the upper atmosphere, the effect upon the blue and violet rays from the sun is very great. The volcanic dust particles are so large that instead of scattering these rays of shorter wave-length, as do the oxygen and nitrogen in our atmosphere, they reflect them back into space and so decrease the amount of light and heat received from the sun. For this reason the general temperature of the earth is lowered by violent volcanic eruptions. Unusually cold periods, that lasted for months, followed the terrible eruption of Krakatoa in 1883 and of Katmai in 1912.

At times when much dust is present in the atmosphere, the sky is a milky white color by day as a result of the reflection of sunlight from the dust particles. Sunrise and sunset colors are then particularly gorgeous, with reds predominating. At such times the blue and violet rays are almost completely shut out, and the red, orange and yellow rays are powerfully diffracted and scattered by the dust particles in the air.

The twilight glow that is visible for some time before sunrise or after sunset is, of course, entirely an atmospheric effect caused by the reflection of sunlight to our eyes from the upper atmosphere, upon which the sun shines, while it is, itself, concealed from our view below the horizon. The atmosphere extends in quantities sufficient to produce twilight to an elevation of about sixty miles.

When all the rays of which sunlight is composed are reflected in equal proportions we get the impression of white light. Dust and haze in the air reflect all rays strongly and give a whitish color to an otherwise blue sky. Brilliant white clouds appear white, because they are reflecting all rays equally. Clouds or portions of clouds appear black when they are in shade or, at times, by contrast with portions that are more strongly illuminated, or when they are moisture-laden and near the point of saturation, when they are absorbing more light than they reflect. At sunrise and sunset, when the light that falls upon the clouds is richest in red and orange and yellow, clouds reflect these colors to our eyes, and we see the brilliant sunset hues which are more intense the more the air is filled with dust and impurities.

The familiar and beautiful phenomenon of the rainbow is produced by refraction, reflection and interference of sunlight by drops of falling water, such as rain or spray. As the ray of sunlight enters the drop of water, which acts as a tiny glass prism, it is refracted or bent from its course and spread out into its spectral colors. Reflection of these rays next takes place (once or twice, as the case may be) from the inside of the drop and a second refraction of the reflected ray takes place as it leaves the drop. The smaller the drops the more brilliant is the rainbow and the richer in color. The most brilliant rainbows are produced by drops between 0.2 and 0.4 millimeters in diameter. In addition to the primary bow, which has a red outer border with a radius of 42°, there is the secondary bow with a radius of about 51° and with colors reversed, the red being on the inner border; the supernumerary bows which are narrow bands of red, or green and red, appear parallel to the primary and secondary bows along the inner side of the primary bow and the outer side of the secondary bow. No rainbow arches ever appear between the primary and secondary bows, and it can be shown in fact, that the illumination between these two bows is at a minimum.

The primary, secondary and supernumerary bows all lie opposite the sun in the direction of the observer's shadow and the observer must stand with his back to the sun in order to see them. The primary and secondary rainbow arches take the form of arcs of circles that have their common center on the line connecting the sun with the observer at a point as far below the horizon in angular distance as the sun is above the horizon. It is, therefore, never possible to see a rainbow arch of more than a semicircle in extent unless the observer is at an elevation above the surrounding country, under which circumstances it might be possible to see a complete circle formed by the rainbow.

The highest and longest arch appears when the sun is on the horizon, and the greater the altitude of the sun the smaller and lower the visible arch. As the angular radius of the primary bow is 42° and of the secondary bow 51° and as the common center of the two circles is always as far below the horizon as the sun is above, it is never possible to see either primary or secondary rainbow when the altitude of the sun is over 51°, or the primary bow when the altitude is over 42°. For this reason rainbows are rarely seen at or near noon in mid-latitudes, since the sun is usually at an elevation of more than 42° at noon, especially in the summer season, which is also the most favorable season for rainbows, owing to the great likelihood of rain and sunshine occurring at the same time.

The light which comes to an observer from the primary bow is once reflected within the drop, and that which comes from the secondary bow is twice reflected within the drop. The sharper and brighter light therefore comes from the primary bow of 42° radius. The space between the two bows is particularly dark, because it can be shown that the drops there do not reflect any light at all.

The rainbow colors are rarely pure or arranged in spectral order, owing to interference of light waves. It is the interference of light waves from different parts of the same drop that produces the bands of alternate maximum and minimum brightness, that lie below the primary bow and beyond the secondary bow. The red or green and red bands of maximum brightness produced thus by interference, are called the supernumerary bows, and they are always found parallel to the primary and secondary bow within the former and above the latter.

The distance of the rainbow from the observer is the distance of the drops that form it. A rainbow may be formed by clouds several miles distant or by the aid of the garden hose on our lawn. No two observers can see exactly the same rainbow because the rainbow arch encircles the surface of a cone whose vertex is at the observer's eye and no two such vertices can exactly coincide. Two observers see rainbows formed by different drops.

Refraction of light by ice-crystals in clouds produces many beautiful color effects, such as halos of various types around sun or moon, vertical light pillars, circumzenithal arcs, and parhelia—"sun-dogs"—or paraselenæ—"moon-dogs"—

which are luminous spots at equal altitudes with sun or moon—one to the left the other to the right, at an angular distance of 22°.

The most usual form of halo is that of 22° radius. This is a luminous ring of light surrounding sun or moon, with the inner edge red and sharply defined and the spectral colors proceeding outward in order; red is frequently the only color visible, the remainder of the ring appearing whitish. Since the halo is produced by refraction of light by ice-crystals, which exist in clouds of a certain type gathering at high altitudes, it is always a very good indicator of an approaching storm.

Coronas are luminous rings showing the spectral colors in the reverse order, that is, with the inner edge blue instead of red. They are usually of very small radius, scarcely two degrees, closely surrounding sun or moon and are produced—not by refraction—but by diffraction or a bending aside of the rays as they pass between—without entering—very small drops of water in clouds. As in the case of refraction, the red rays are turned from their course the least and the violet rays the most.

Many of these phenomena—halos, luminous spots, vertical pillars and arcs of light may, at times, be seen simultaneously, when clouds of ice-crystals are forming around the sun or moon. They then present a very complex and beautiful outline of luminous circles, arches and pillars that have a mysterious and almost startling appearance when the cause is not clearly understood.

We have found then that sunlight is made up of rays of many different wave-lengths and colors and that the atmosphere acts upon these rays in various ways. It reflects them or turns them back on their course; it refracts them as they pass through the gases of which the atmosphere consists, or through the water-vapor and ice-crystals suspended in it, thus sifting out and dispersing the rays of different colors and wave-lengths and producing beautiful color effects; it diffracts them or bends them aside as they pass between the fine dust particles and small drops of water in the air, again sifting out the rays of different colors and producing color effects similar to those produced by refraction; it also scatters and disperses, through the action of the molecules of oxygen and nitrogen in the upper strata, the blue and violet rays of shorter wave-length and thus produces the blue color and brightness of the sky; it produces beautifully colored auroral streamers and curtains and rays of light through the electrical discharges resulting when the rarefied gases in the upper air are bombarded by electrified particles shot forth from the sun.

It is our atmosphere, then, that we have to thank for all these beautiful displays of color that delight our eyes and give pleasure to our existence, as well as for the very fact of our existence upon a planet that without its presence would be an uninhabitable waste, covered only with barren rocks.

## CHAPTER XXVI. KEEPING TRACK OF THE MOON.

Of all celestial objects the nearest and most familiar is our satellite, the moon. Yet the mistakes and blunders that otherwise intelligent persons frequently make when they refer to the various aspects of the moon are quite unbelievable.

Who has not read in classics or in popular fiction of crescent moons riding high in midnight skies, of full moons rising above western cliffs or setting beyond eastern lakes? Who has not seen the moon drawn in impossible positions with horns pointing toward the horizon, or a twinkling star shining through an apparently transparent moon?

Careful observation of the moon in all its various phases and at different seasons is the best method to be used in acquiring a knowledge of the elementary facts regarding the motion of the moon through the heavens from day to day, but that requires that one be up often after midnight and in the early hours preceding dawn and so it is that we feel so hazy in regard to what happens to the moon after it has passed the full.

A few fundamental rules can be easily acquired, however, and these will enable us to locate the moon in the right quarter of the heavens at any time of the day or night when we know its phase and the approximate position of the sun at the same instant, and thus we may avoid some of the most obvious blunders that are made in dealing with the general aspect of the moon at any given time.

As can be verified by direct observation, the moon is always moving continually eastward. Since it makes a complete revolution around the earth from new moon back to new moon again in a little less than thirty days, it passes over about twelve degrees a day (360° divided by 30), on the average, or one-half a degree an hour, which is about the angular extent of its own diameter. Therefore every hour the moon moves eastward a distance equal to its own diameter. This is of course only approximate as the moon moves more rapidly in some parts of its orbit than in others.

In addition to its real eastward motion the moon shares the apparent daily westward motion of all celestial objects which is due to the daily rotation of the earth on its axis in the opposite direction. That is, the moon, as well as the sun, stars and planets, rises in the east and sets in the west daily. On account of its continuous eastward motion, however, the moon rises later every night, on the average about fifty minutes, though the amount of this daily retardation of moonrise varies from less than half an hour to considerably over an hour at different seasons of the year and in different latitudes. In the course of a month then the moon has risen at all hours of the day and night and set at all hours of the day and night.

It might seem unnecessary to emphasize the fact that the moon always rises in the east were it not that the astronomer occasionally meets the man who insists that he has at times seen the moon rise in the west.

To be sure the new crescent moon first becomes visible above the western horizon shortly after sunset though it rises in the east the morning of the same day shortly after sunrise. As is also true of the sun the exact point on the horizon where the moon rises or sets varies from day to day and from season to season. In one month the moon passes over very nearly the same path through the heavens that the sun does in one year, for the moon's path is inclined only five degrees to the ecliptic or apparent path of the sun through the heavens. It can never pass more than 28½° (23½° + 5°) south of the celestial equator, nor more than 28½° north of it. It has a slightly greater range in altitude than the sun, therefore. North of 28½° north latitude it always crosses the meridian south of the zenith and below 28½° south latitude it crosses the meridian north of the zenith. In tropical regions the moon sometimes passes north of the zenith, sometimes south, or again directly through the zenith.

Since the full moon is always diametrically opposite to the sun it passes over nearly the same part of the heavens that the sun did six months before. In winter then when the sun is south of the equator the moon "rides high" at night north of the equator and, vice versa, in summer when the sun is north of the equator the full moon "rides low" south of the equator. In winter then we have more hours of moonlight than we have in summer. This may be of no great advantage in mid-latitudes but we may imagine what a boon it is to the inhabitants of the Arctic and Antarctic regions to have the friendly moon above the horizon during the long winter months when the sun is never seen for days at a time.

At time of "new" moon the moon lies directly between us and the sun, but ordinarily passes just to the north or south of the sun since its orbit is inclined five degrees to the ecliptic or plane of the earth's orbit. If the moon's path lay exactly in the ecliptic we would have an eclipse of the sun every month at new moon and an eclipse of the moon two weeks later at full moon. Now the moon crosses the ecliptic twice a month, the points of crossing being called the nodes of its orbit, but only twice a year is the moon nearly enough in line with the sun at the time it crosses to cause eclipses. Every year, then, there are two "eclipse seasons," separated by intervals of six months, when the moon is in line with the sun at or close to the point where it crosses the ecliptic; then and only then can solar and lunar eclipses occur. The solar eclipses, of course, will occur when the moon is new, that is, when the moon passes directly between the earth and the sun and throws its shadow over the earth; and the lunar eclipses two weeks later when the earth passes between the sun and moon and throws its shadow over the face of the moon.

Probably there is no astronomical subject that has been more generally misunderstood than that of solar and lunar eclipses. It is well to remember that solar eclipses can only occur at time of new moon and lunar eclipses only at the time of full moon; and at the time of eclipses, whether lunar or solar, the moon is at or near its nodes, the points where its orbit crosses the ecliptic. There are always at least two solar eclipses every year and there may be as many as five. There are years when there are no lunar eclipses, though ordinarily both solar and lunar eclipses occur every year, some partial others total.

The moon shines only by reflected sunlight. It is of itself a solid, dark body with its day surface intensely hot and its night surface intensely cold, a world of extreme temperatures.

At new moon all of the night side of the moon is turned toward us, at full moon all of the day side. At other phases we see part of the day side and part of the night side and the illuminated side of the moon is always the side that is towards the sun. Failure to observe this simple rule leads to many grievous blunders in depicting the moon.

At the time of new moon the moon, moving continually eastward, passes north or south of the sun from west to east except when it passes directly in front of the sun, causing eclipses. A day or so later the waxing crescent moon or the "new moon," as it is popularly called, becomes visible low in the west immediately after sunset. The moon is now east of the sun and will remain east of the sun until the time of full moon. During the period from new moon to full moon it will, therefore, rise after the sun and set after the sun. The waxing crescent moon will not be visible in the morning hours because, inasmuch as it rises after the sun, it is lost to view in the sun's brilliant rays. Nevertheless, it follows the sun across the sky and becomes visible in the west as soon as the sun has disappeared below the western horizon. The thin illuminated crescent has its horns or cusps turned away from the point where the sun has set. The horns of the crescent can never point toward the horizon since the illuminated side of the moon is always turned toward the sun whether the sun is above or below our horizon.

As hour by hour and day by day the moon draws farther eastward and increases its angular distance from the sun, more and more of the illuminated side becomes visible; the crescent increases in width and area and the moon appears higher in the western sky each night at sunset.

Usually about seven and a fraction days after the date of new moon the moon completes the first quarter of its revolution around the earth. The period from one phase to the next is variable and irregular, being sometimes less than seven days and at other times more than eight days, since the moon does not move at a uniform rate in different parts of its orbit.

When the moon has completed the first quarter of a revolution it is ninety degrees east of the sun and presents the phase known as "half-moon" since half of the surface that is turned toward the earth is illuminated and half is in darkness. It is said to be "at the first quarter." The illuminated half is of course the western half because the sun is to the west of the moon. The half moon is near the meridian at sunset and sets near midnight. Up to the first quarter, then, the moon is a crescent in the western sky during the first part of the night and should never be represented as east of the meridian or near the meridian at midnight.

After the moon has passed the first quarter and before it is full more than half of the side turned toward the earth is illuminated and it is in the "gibbous" phase. It is still the western limb that is fully illuminated. The moon is now east of the meridian at sunset and it crosses the meridian before midnight and sets before sunrise. All who are abroad during the first half of the night find this phase of the moon more favorable to them than the gibbous phase following full moon.

The moon now being above the horizon at sunset is visible continuously from sunset to midnight but sets some time during the second half of the night, while the full moon shines throughout the night, rising in the east at sunset and setting in the west at sunrise.

When the moon is full it is 180° east, or west, of the sun and so both its eastern and western limbs are perfectly illuminated. After the full the moon goes through its phases in reverse order, being first gibbous, then a half-moon once more, and lastly a waning crescent.

It is now west instead of east of the sun and so it is the eastern limb that is fully illuminated by the sun. Being west of the sun it will now rise before the sun and set before the sun, the interval decreasing each day as the moon draws in toward the sun once more.

The gibbous phase preceding full moon is favorable to all abroad before midnight but the gibbous phase following full moon is more favorable to those who are abroad after midnight, for from full moon to last quarter the moon is below the horizon at sunset, and of course, is rising later and later each night, while at sunrise it is still above the horizon, appearing each day higher and higher above the western horizon at sunrise as it approaches the third or last quarter.

When it has reached this point it is once more a half-moon, though now it is the eastern half instead of the western half of the disk that is fully illuminated. The moon is 90° west of the sun at third quarter and from this phase to the phase of new moon it is a crescent once more, but now a waning instead of a waxing crescent. It appears east of the meridian before sunrise and as the crescent grows thinner it draws nearer and nearer to the eastern horizon and the rising sun. As with the waxing crescent moon the horns are turned away from the horizon. The waning crescent moon is always to be looked for east of the meridian and to be associated with the rising sun, while the waxing crescent moon is to be looked for west of the meridian and associated with the setting sun. Neither the waxing nor the waning crescent moon will be visible during the midnight hours.

As the waning crescent moon grows thinner and draws in closer to the sun each successive night, its time of rising precedes that of the sun by an ever-decreasing interval until finally the crescent disappears from view in the eastern sky; the next day we see no crescent either in the eastern or western skies—the moon is once more in conjunction with the sun and "new." One revolution of the moon about the earth with respect to the sun has been completed and a day or so later we may look for a new crescent moon in the western sky after sunset.

## CHAPTER XXVII. THE MOTIONS OF THE HEAVENLY BODIES.

About three hundred and twenty years ago Giordano Bruno was burned at the stake for his audacity in believing in the existence of other worlds. A few decades later the famous astronomer Galileo was forced to publicly recant his belief that the earth moved. Yet the truth could not long be suppressed by such means, and since those dark days man's advance in knowledge has been so rapid that it seems to us today in this wonderful age of scientific discovery almost inconceivable that man ever believed that the earth, a tiny planet of a vast solar system, was "the hub of the universe," the fixed and immovable center about which revolved all the heavenly bodies. Very reluctantly, however, and with bitter feeling, but in the light of overwhelming evidence man finally gave up his long-cherished idea of terrestrial importance, and when finally forced to move his fixed center of the universe, he moved it only so far as the comparatively nearby sun.

This center he then regarded as fixed in space and also held to his belief that the stars, set in an imaginary celestial sphere, were immovable in space as well, and all at the same distance from the sun. So, scarcely two hundred years ago we find that the astronomer Bradley was endeavoring to measure this common distance of the "fixed stars." Though he failed in this attempt he made the important discovery that the observed positions of the stars are not their true positions, owing to the fact that the velocity of light is not infinite but takes a definite finite interval of time to travel a given distance. As a result the stars always appear displaced in the direction of the earth's motion around the sun, the amount of the displacement depending upon the velocity of the earth in its orbit and the velocity of light. This "aberration of light," as it is called, furnished additional proof that the earth revolves about the sun and was one more nail driven into the coffin of the old Ptolemaic theory that the earth was the center of the universe. Bradley also discovered that the positions of the stars were affected by the wabbling of the earth's axis, called its "nutation."

Although in the days of Bradley neither the methods of observation nor the instruments were sufficiently accurate to show the minute shifts in the positions of the stars that reveal the individual motions of the stars and the distances of those nearest to us, yet the discovery of the two large displacements in the positions of all the stars, due to the aberration of light and the nodding of the earth's axis were of the greatest value, for they were a necessary step in the direction of the precise measurements of modern times. It is only through measurements of the greatest refinement and accuracy that it is possible to detect the motions and distances of the stars and to discover the wonderful truths about the nature and structure of the universe that they are revealing to us today.

After unsuccessful attempts extending over several centuries the distance of one of the nearest stars, the faint 61 Cygni, as it is catalogued, was finally determined by the astronomer Bessel in the year 1838.

This star is about ten light-years distant from the earth, which places it about six hundred and thirty thousand times farther away from us than the sun; that is, we would have to travel six hundred and thirty thousand times the distance from the earth to the sun to reach this very close stellar neighbor, 61 Cygni. The nearest of all the stars, Alpha Centauri, is over two hundred and seventy thousand times the distance from the earth to the sun. It is, therefore, little wonder that the early astronomers believed that the stars were fixed in space since even the nearest is so far away that, viewed from opposite points in the earth's orbit, its apparent change in position due to our actual change in position of 186,000,000 miles, amounts to only one and a half seconds of arc. Two stars separated by one hundred and sixty times this angular distance might possibly be glimpsed as two distinct stars by a person with good eyesight, though to most of us they would appear as one star. Upon the measurement of such minute angles depended a knowledge of the distances of the nearest stars.

It is to Sir William Herschel that we owe the discovery, more than a hundred years ago, of the motion of the sun through the universe. From the consideration of a long series of observations of the positions of the stars this famous astronomer discovered that the stars in the direction of the constellation Hercules were separated by much greater angular distances than the stars diametrically opposite in the heavens. In other words, the stars were spreading apart in one portion of the heavens and crowding together in the opposite direction and he rightly interpreted this to mean that the sun was moving in the direction of the constellation of Hercules. It was not until the spectroscope was applied to the study of the heavens in the latter part of the nineteenth century that the amount of this motion of the sun was found to be about twelve and a half miles per second, or four times the distance from the earth to the sun in a year.

It is to Sir William Herschel that we owe also the discovery of binary systems of stars in which two stars swing around a point between them called their center of gravity.

Our first conception of the immensity and grandeur of the universe dates from the time of the older Herschel only a century or so ago. The mysterious nebulæ and star clusters were then discovered, the wonders of the Milky Way were explored, and a new planet and satellites in our own solar system were discovered. It was found that the sun and the stars as well as the planets were in motion. Neither sun nor earth could be regarded any longer as a fixed point in the universe.

With the application of the spectroscope to the study of the heavens toward the end of the nineteenth century the key to a treasure-house of knowledge was placed in the hands of the astronomers of modern times and as a result we are now learning more, in a few decades, about the wonders and mysteries of the heavens than was granted to man to learn in centuries of earlier endeavor. Yet it is the feeling of the astronomer of today that he is only standing on the threshold of knowledge and that the greatest of all discoveries, that of the nature of matter and of time and space is yet to be made.

It is the spectroscope that tells us so many wonderful facts about the motions of the stars, nebulæ and star clusters. It tells us also practically all we know about the physical condition of our own sun and of the other suns of the universe, their temperature and age, and the peculiarities of their atmospheres.

Some of the most important astronomical discoveries that have been made in the past few years have to do with the distribution and velocities of the heavenly bodies as revealed by the spectroscope.

It has been found, with the aid of the spectroscope, that the most slowly moving of all stars are the extremely hot bluish Orion stars with an average velocity of eight miles per second, while the most rapidly moving stars are the deep-red stars with an average velocity of twenty-one miles per second, and there is in all cases a relationship existing between the color, or spectrum, of a star and its velocity. The reason for this connection between the two still remains undiscovered.

The spectroscope has also told us some astonishing facts in recent years about the velocities of the spiral nebulæ.

It is now known that these mysterious objects are moving with the tremendous average velocity of four hundred and eighty miles per second, which exceeds the average velocity of the stars fully twenty-five fold. They possess, moreover, internal motions of rotation that are almost as high as their velocities through space. It is now generally believed that spiral nebulæ are far distant objects of enormous size and mass, exterior to our own system of stars and similar to it in form.

In place of the universe of the "fixed stars" and the immovable sun or earth of a few centuries ago we find that modern astronomical discovery is substituting a universe of inconceivable grandeur and immensity in a state of ceaseless flux and change.

Our earth—an atom spinning about on its axis and revolving rapidly around a huge sun that is equal in volume to more than a million earths—is carried onward with this sun through a vast universe of suns.

Only an average-sized star among several hundred million other stars is this huge sun of ours, moving with its planet family through the regions of the Milky Way, where are to be found not only moving clusters and groups of stars, speeding along their way in obedience to the laws of motion of the system to which they belong, but also strangely formed nebulæ covering vast stretches of space, whirling and seething internally and shining with mysterious light, and still other stretches of dark obscuring matter shutting off the rays of suns beyond.

The extent and form of this enormous system of stars and nebulæ and the laws that govern the motions of its individual members are among the problems that the astronomers of today are attempting to solve. On both sides of these regions of the Milky Way, wherein lies our own solar system, lie other vast systems, such as the globular star clusters, composed of thousands, possibly hundreds of thousands, of suns; the Magellanic clouds, which resemble detached portions of the Milky Way, and, probably, the much discussed spiral nebulæ, possible "island universes" similar to our own.

We have come far in the past three hundred years from the conception of an immovable earth at the center of the universe to this awe-inspiring conception of the universe that we have today, which is based upon modern astronomical discoveries.

Whatever may be discovered in the future in regard to the form and extent of the universe the idea of a fixed and immovable center either within the solar system or among the stars beyond has gone from the minds of men at last.

Not more than a generation ago a survival of the old idea of a fixed center was seen in the belief that Alcyone, in the Pleiades was a "central sun" about which all the stars revolved. It is now well known that the Pleiades form a moving star cluster. Alcyone is therefore drifting slowly onward through the universe and the idea of a fixed and immovable center to which man may anchor his ideas is drifting away also. There are, it is true, local centers of systems, such, for instance, as the sun occupies in the solar system or some group of stars may occupy in the stellar system to which our sun belongs, yet as a whole these systems move on and their centers with them. There is no evidence today that any absolutely immovable point exists in the heavens.

No celestial object has been found to be without the attribute of motion, not only motion onward through the universe, but also rotational motion about an axis of the body. The planets rotate on their axes as well as revolve about the sun, and the sun also turns on its axis as it moves onward through space. This rotational motion is also found in the nebulæ and star clusters as well as in the stars and planets. No object in the heavens is known to be without it. Even the slowly drifting Orion nebula possesses a rapid internal velocity of rotation. There is no such thing as a body absolutely at rest in the universe.

# CHAPTER XXVIII. THE EVOLUTION OF THE STARS—FROM RED GIANTS TO RED DWARFS.

The most casual of star-gazers is aware that the stars differ one from another in color and in brightness. There are red stars, yellow stars, white stars and bluish-white stars. There are the brilliant stars of first magnitude such as Vega, Capella and Antares, and there are, on the other hand, stars so faint that they can barely be glimpsed with the most powerful telescopes.

In general the most brilliant stars are the nearest and the faintest stars are the most distant, but there are many exceptions to the rule, since there are stars that appear faint even when comparatively near because they are small and shine with a feeble light. Such a star is the faint, sixth-magnitude star, 61 Cygni, one of the nearest of all the stars. Again, there are stars in far-distant clusters visible only in powerful telescopes that in actual brightness exceed our own sun several thousand times and in volume several million times. A star the size of the sun would be invisible in the most powerful telescope in existence if it were at the distance of many stars in the Milky Way or globular star clusters.

Stars differ in color because they differ in temperature. We are all aware of the fact that a piece of iron when heated first glows a deep red, then appears yellowish in color and finally attains to white heat. It is the same among the stars. The red stars are the coolest of all the stars and the bluish-white stars are the hottest of all the stars, while intermediate between them in temperature come the yellow and the white stars.

Now as the biologist and the geologist see in this world of ours signs of evolution, or gradual development and change from the simple to the more complex forms, and of growth and decay, so the astronomer sees among the stars signs of a continuous, progressive development from one type of star to another. Stars share in the general evolution that is the law of the universe, and are born, reach the height of their development, decline to old age and die.

Within the past few years important astronomical discoveries have been made that show the true order of this evolution of the stars. It was believed not so long ago that the blue-white helium stars—the type B stars the astronomers called them, or the Orion stars, since there are so many stars of this type in the constellation of Orion—were not only the hottest but also the youngest of the stars and that they represented the first stage in the development of a star from a primitive gaseous nebula such as the Great Orion Nebula. It is now known that these brilliant, hot helium stars represent the peak of development of the most massive of all the stars and not the first stages in the development.

A star, it is now known, comes into existence as a giant, reddish star of enormous size and of a density only about one-thousandth that of the earth's atmosphere at sea-level. It is inconceivably tenuous or rare, and its temperature is comparatively low, about 3,000° Centigrade or less. It is not evolved from the luminous, gaseous nebulæ because red stars are never found associated with the gaseous nebulæ, as are the blue-white stars. The origin of these red giant stars is uncertain, but it is possible that they may be gradually evolved in some manner from the dark clouds of obscuring matter or dark nebulæ that exists so abundantly in the heavens.

In the next stage of its development the deep-red giant star increases in temperature as it contracts under the action of gravitation and its color gradually changes from red to yellow. Its density increases slightly and its volume decreases. It is now a yellow giant star. As the evolution progresses in the course of ages the star continues to contract, its temperature increases greatly as does also its density and it continues to decrease in volume. It is now a brilliant white star, a hydrogen star, so called because its spectrum is chiefly characterized by the lines of hydrogen.

As the star contracts under the gravitation of its parts and increases in temperature and density there comes more and more into play an important factor that has a great effect upon its future development. This is light-pressure or radiation pressure which acts in opposition to gravity and exerts a strong outward pressure upon matter within the depths of the star, tending to push it outward from the center where the temperature is greatest and the light is most intense. It is a most interesting fact that if the mass of a star, that is the quantity of matter that it contains, exceeds a certain value the pressure of light or radiation within it overbalances the gravitational attraction that draws matter towards its center and the star disintegrates or ceases to exist as a star. This accounts for the fact that the stars differ very little among themselves in the quantity of matter that they contain, that is, in their masses, though they may differ enormously in size. Stars that exceed a certain mass will become unstable and this may account for the association of luminous nebulæ with the hottest of all stars, the nebulæ possibly being puffed off from the surfaces of these stars under the action of radiation pressure.

After a star has reached the height of its development as a bluish-white helium star with a temperature of something like 10,000° Centigrade and a density about one-tenth that of the sun, it begins to lose heat and to cool gradually though it continues to contract and increase in density.

It is now on the descending scale of evolution and is to be counted among the dwarfs instead of the giants. A brilliant blue-white helium or Orion star is about one hundred times more luminous than the sun, and its diameter is about ten times that of the sun.

Our own sun, we find, is on the descending scale of stellar evolution. It is a yellow dwarf star of temperature about 6,000° Centigrade and density one and one-fourth that of water, which is probably about as great a density as is attained by any star since even the non-luminous planets Jupiter and Saturn have lower densities than the sun.

The last stage in the development of a star is represented by the dwarf red star of high density and low temperature. The diameter of the dwarf red star probably averages about five hundred thousand miles and its temperature is 3,000° Centigrade or less. After this we have the extinct stars, similar possibly to our planet Jupiter, though considerably larger, with a dense gaseous atmosphere and a certain degree of internal heat.

We have traced the evolution of a star from a red giant to a red dwarf through the intermediate stages from yellow giant to a giant helium star with increasing temperature and thence to yellow dwarf and red dwarf as the temperature decreases. Only the most massive stars pass through this entire chain of evolution. Stars of small mass never attain to the splendor of brilliant blue-white helium stars, but begin to decrease in temperature and brightness before this stage is reached.

The time required for the evolution of a star from red giant to red dwarf is not known, but it must be very great. The age of the earth, which is probably equal to that of the solar system, is estimated as something like one thousand million years. It is probable that the average life of a star far exceeds this limit.

# CHAPTER XXIX. DOUBLE AND MULTIPLE STARS.

The plan of the solar system which consists of a central sun encircled by satellites that are far inferior to their luminary in size, and that move about it in orbits that are almost perfect circles, is not the only, nor possibly, even the most general one in the universe.

Sweeping the heavens with powerful telescopes one is astonished to find that myriads of stars can be separated into two or more physically connected suns that are often, moreover, of exquisitely tinted and contrasting shades. Green and red, orange and blue, white and golden or white and blue pairs exist in profusion, and strange to say there are well-authenticated instances of color changes taking place temporarily within the same system. A pair of white stars has been known to change within a few decades, first to golden yellow and bluish green and then to orange and green. The famous pair catalogued as "95 Herculis" was noted to change from green and red to a palish yellow and back to the original strongly contrasting hues within the course of a single year, while at another time they appeared to be a perfectly white pair. At the present time both of these stars are decidedly yellowish in color. Such changes in hue are probably due to temporary disturbances in the atmospheres of the stars, possibly of an electrical nature or to sudden or unusual outbursts of activity, concerning the origin of which we are as much in doubt as we are of the cause of the sun-spot cycle and periodic variation in the intensity of radiation of our own sun. Temporary changes in the color of the components of double star systems sometimes take place when the two stars approach their "periastron" or point of nearest approach. Owing to the great eccentricity of the orbits of double stars, such stars are anywhere from twice to nineteen times as near to each other at periastron as they are at "apastron," or point of greatest departure. Such great changes in the relative distances of two physically connected suns would produce marked changes in the intensity of the tides raised upon each of them by their mutual gravitational attraction and unusual outbursts of gases or electrical excitement in the atmospheres of the stars might cause very noticeable changes in the color of these stars as they drew nearer to each other, which would subside as they receded toward apastron.

In addition to "visual" double or multiple stars, there exists a very extensive class of stars known as "spectroscopic binaries," in which the two components are so close to each other that even the most powerful telescopes cannot divide them. It is only from the shifting of the lines of their overlapping spectra, caused by their alternate motion toward and from the earth as they revolve about their common center of gravity, that their duplex nature is revealed to us.

In some instances one member of the system is so faint that its spectrum is not visible and its presence is disclosed only by the shifting of the lines of the bright star.

According to Doppler's Law, when a star is approaching the earth the lines of its spectrum shift toward the blue end of the spectrum, and when the star is receding from the earth the lines are shifted toward the red end of the spectrum. The amount of this shift can be very accurately measured, and gives the relative velocities of the stars in their orbits directly in miles per second. Knowing in addition, by observation, the period of mutual revolution of the stars, it is possible to find the dimensions of these spectroscopic binary systems compared to our own solar system, and also the masses of the stars compared to the mass of our own sun. If the spectrum of the fainter star is not visible, only the velocity of the brighter star with respect to the center of gravity of the system can be found and the mass found for the system comes out too small. In such cases we can obtain only a lower limit for the mass of the system. Then, too, it must be remembered that these systems of stars lie at all angles with reference to our line of sight, and so we rarely see the orbits in their true form. The measured velocities are as a result smaller than the true velocities, and on the average amount to only sixty per cent. of the true orbital velocities. The calculated masses of spectroscopic binary stars are, therefore, in general only about sixty per cent. of the true masses. It has been found from calculating the masses of a number of binary systems, that the combined masses of the stars in these systems do not differ very greatly among themselves, nor as compared to our own sun, though in light-giving power these stars may differ hundreds, thousands, even millions of times. For instance, there are stars that give only one ten-thousandth part of the light of our own sun, and other stars that give ten thousand times as much light as the sun. Moreover, there are many instances of physically connected stars differing thousands of times in luminosity, though in mass, or quantity of matter found in the stars, they differ only two or three times. Why this is so remains one of the great mysteries of the heavens, and makes it extremely difficult to give any satisfactory theory of the origin of double-star systems. It has never been explained satisfactorily why of two suns physically connected and, therefore, presumably originating at the same time, one should be radiating with the greatest intensity, while the other is practically an extinct sun, in spite of the fact that the quantity of matter in the two bodies differs but slightly.

In a few systems the plane in which the stars revolve passes so nearly through the earth that the two stars temporarily eclipse one another during each revolution. Such systems are called eclipsing binaries. To such a system belongs the famous

Algol. Its light waxes and wanes periodically with the greatest punctuality in a period of 2d 20h 48.9m, owing to its temporary eclipse by a very large but extremely faint attendant sun. The diameter of the faint star is slightly greater than the diameter of the bright star which is about one million miles in extent. The distance between the centers of the stars is only about three million miles, which brings their surfaces within two million miles of each other. The masses of the two stars are in the ratio of two to one, the brighter and more massive star being about half as heavy as our own sun, though its density is only about two-tenths that of the sun. The density of the fainter star is still less, being only about half that of the brighter star. Very low density of both components and extreme faintness of one member compared to the other, appears to be a very general characteristic of closely associated eclipsing and spectroscopic binary stars. Among the extremely hot and brilliant helium and hydrogen stars, spectroscopic binaries exist in great numbers. In fact, among these types there appear to be as many binary and multiple systems as there are systems of isolated suns. Sometimes these close binary stars are egg-shaped or oval and revolve rapidly almost in contact about their common center of gravity. Inhabitants of satellites of such a system would see in their heavens the, to us, strange and startling phenomenon of two suns, each equal to our own or even greater in size, whirling rapidly about each other and separated by a space comparable in extent to their own diameters. Eclipses in such a system would be of daily occurrence, and, if one star were dark, would produce for the satellite world the same effect of alternate day and night that results from axial rotation of a satellite. The two hemispheres of the faint companion sun would be very unequally illuminated owing to the fact that the side turned toward the brilliant sun would always reflect its neighbor's brightness in addition to shining with its own comparatively feeble inherent light, while the opposite hemisphere would shine only by its own dim light, and would, therefore, be in comparative darkness.

The spectroscopic binaries generally revolve closely and rapidly about their common center of gravity; there are to be found, on the other hand, among the wider visual doubles, many systems wherein the components are separated by distances comparable to the distances of the outer planets, Saturn, Uranus and Neptune, from the sun. It is evident that the individual stars of such binary systems could not possibly be encircled by any such extensive system of satellites as attends our own sun, though satellites such as our own planet Earth, or the inferior planets Mercury and Venus, might conceivably encircle the individual components of such binary systems at distances not greater than that of the earth from the sun. No planet could safely exist at a much greater distance from one of these suns without being subject to most dangerous perturbations and disruptive tidal forces arising from the vicinity of the second sun. Granted that planets might encircle one of these suns at a distance approximating that of Venus or our own planet from the sun, the inhabitants of such worlds would behold the strange phenomenon of two suns in the heavens, not almost in contact as in spectroscopic binary systems, but at one time comparatively near and again in opposite portions of the heavens as is the case with the sun and moon in our own heavens. As the planet advanced in its orbit about the ruling sun, the secondary sun would be visible at first by day and again by night. If the two suns were of contrasting hues, as, for instance, green and red, there might appear in the nearby heavens at a distance of one hundred million miles or so a magnificent sun of deep reddish hue, equal to or surpassing our own in splendor, while in a far distant part of the sky, at a distance as great as that which separates us from the planet Saturn, might appear a rival sun of greenish hue, smaller and fainter, but nevertheless, hot and extremely brilliant and capable of exerting through its great gravitational attraction a most disturbing effect upon the motion of the planet of its neighbor. At times the rays of the two suns, red and green, would combine to produce a day characterized by terrific heat and intense illumination. Again the green orb would rise in the east as the red sun set in the west and night would be turned into a weird, dimly-lighted day by the greenish rays of the secondary sun. Compared to the wonders and beauties of the heavens in such a system, our own well-regulated and orderly planet family, undisturbed by the exciting proximity of a rival sun, seems to pale into insignificance. Yet we have every good reason to be content with the ordering of affairs within our own solar system, and to feel relief rather than regret at the absence of a secondary sun. In a planet world revolving about one member of a double star system, we may imagine the dread rather than pleasure with which the periodic near-approach of a rival sun would be hailed, and even the possible hurried migration from exposed to sheltered portions of the planetary world to escape the rapidly increasing heat and intensity of light from the approaching sun. In such systems the coming and going of the seasons might indeed be a matter of life and death to the inhabitants of satellite worlds!

Within our solar system the masses of the planets are practically negligible compared to the mass of the sun, and it is for this reason that they appear to revolve about the center of the sun. As a matter of fact, no body in the universe revolves about the exact geometrical center of another body, but two mutually attracting bodies revolve in orbits about their common center of gravity, which always lies between the two bodies on the line connecting them and at a distance from each of them that is in inverse proportion to the mass of the body. The moon does not revolve about the center of the earth, but about the center of gravity of the earth and moon, which lies on the line connecting the two bodies and at a distance from the earth's center that is one eighty-first of the distance from the center of the earth to the center of the moon, since this represents the ratio of the masses of the two bodies. This center of gravity of the earth and moon, lies, then, about two thousand miles from the earth's center, and about this point both earth and moon trace out orbits of revolution that are identical in form and differ only in size. In the same way each of the planets of the solar system revolves about the center of gravity of itself and the sun, but the mass of the sun is so far in excess of the combined masses of all the planets that we may consider, for all practical purposes, that the planets revolve about the sun's center, the center of gravity of the system being within the sun, just as the center of gravity of the earth and moon is within the earth.

Prof. T. J. J. See found from the investigation of forty binary star orbits that the average eccentricity of double star orbits is twelve times as great as the average eccentricity of a planetary orbit, and that the masses of the component suns never differ very greatly. The center of gravity of a binary system, therefore, lies at a great distance from the centers of the stars, and about this point, as a focus, the stars move in orbits that are exactly similar in form but differ in size in inverse proportion to the ratio of the masses. Since the orbits of binaries are, moreover, very highly eccentric, the two suns are, as we have said, anywhere from two to nineteen times nearer to each other at periastron than they are at "apastron."

We have spoken so far only of systems of two associated suns, but many systems exist in which three or more sun-like bodies are in revolution about a common center of gravity. Frequently two fairly close suns are in revolution about a common center of gravity, in a period, say, of fifty or sixty years, while a third sun revolves at a comparatively great distance about the center of gravity of itself and the first pair in a period of several hundred years. Or possibly the third sun also possesses a close attendant and the two pairs revolve in a period of great length about a common center of gravity.

Such, for instance, are the systems of Zeta Cancri and Epsilon Lyræ. In the former system the closer components revolve rapidly about their center of gravity in a period of about sixty years, while the remote companion shows irregularities in its motion that indicate that it is revolving about a dark body in a period of seventeen and a half years, while the two together are revolving very slowly in a period of six or seven centuries, about a common center of gravity with the first pair in a retrograde direction.

The wider pair of Epsilon Lyræ is a naked-eye double for it can be seen as a double star by a keen eye, while even a three-inch telescope will separate each of the components into a double star. So extensive is this system that the periods of revolution of the closer components occupy several centuries, one pair appearing to revolve about twice as rapidly as the other, while the period of revolution of the two pairs about a common center is probably a matter of thousands of years. The gap that separates the two pairs may be so great that light requires months to cross it.

These multiple systems are by no means exceptional. They are to be found in profusion among the brilliant Orion stars. They have been referred to as "knots" of stars and it has been suggested that they may have originated as local condensations in one vast nebulous tract. A system of only two components appears to be the exception rather than the rule, groups of several connected suns being more numerous than single pairs.

In all of these double and multiple systems there exists the possibility of minute satellites, such as our own earth, in attendance upon some one component of the system. Such tiny bodies shining only by reflected light from a nearby brilliant sun would be hopelessly invisible in the most powerful telescope.

We can only assume that it is far more reasonable to believe in than to disprove the existence of such satellites.

Our own solar system, then, represents neither in its mechanical nor physical features, the only possibilities for the maintenance of life; it can neither be considered a unique form, nor even the most generally prevalent form in the universe.

# CHAPTER XXX. ASTRONOMICAL DISTANCES.

The grandeur of the scale upon which the visible universe is fashioned lies almost beyond human comprehension. In measuring the vast extent of our own solar system, which is but a single unit in the system of the stars, we may have recourse to some earthly standard of measurement, such as the mile. But when we desire to express in terms of units that can be grasped by our imagination, the distances of the stars that lie far, far beyond, we find that all ordinary standards of measurement become utterly inadequate for our purpose. In the measurement of celestial distances within the solar system the unit employed is either the familiar mile or kilometer or the "astronomical unit," which is the mean distance from the earth to the sun (ninety-two million nine hundred thousand miles in round numbers).

In the measurement of distances beyond the solar system the unit employed is either the light-year or more recently the parsec, which is rapidly replacing the light-year among astronomers. A "light-year" is the distance that light, with its finite but almost unimaginable velocity of one hundred and eighty-six thousand miles per second, travels in a year. It is equal in round numbers to sixty-three thousand times the distance from the earth to the sun or approximately six thousand billions of miles. The parsec is equal to three and twenty-six hundredths (3.26) light-years, and it is approximately two hundred thousand times the distance from the earth to the sun. It is "the distance of a star with the parallax of a second," a fact which its name, parsec, conveys to us. In other words, at the distance of one parsec the distance from the earth to the sun, "the astronomical unit," would subtend an angle equal to one second of an arc. This angle is spoken of as the parallax of the star. The larger the parallax, that is, the larger the angle the astronomical unit or radius of the earth's orbit subtends, viewed from the star, the nearer the star is to us. The fact that there is no known star within one parsec, or three and twenty-six hundredths light-years, of the sun shows the immensity of the scale of the universe of stars.

Before considering the distances of the stars and the extent of the sidereal system of which our sun and his satellites form a part, let us undertake to express the distance of the sun, moon and planets from the earth and the extent of the solar system in terms with which we are familiar.

The nearest to the earth of all celestial bodies is its satellite, the moon. So near is the moon that if we should make on some great plain a model of the solar system in which the astronomical unit, the distance from earth to sun, would be four hundred feet, the distance between the earth and moon would be only one foot. On the same scale the most distant planet Neptune would be two and one-quarter miles away.

Granted that it were possible to escape the earth's gravitational bonds and to travel by our swiftest means of conveyance, the airplane, through interplanetary space, let us consider how long it would take us to reach the moon, sun and planets if our speed were maintained at a uniform rate of two hundred miles an hour. An airplane traveling at this rate would circumnavigate the earth in a little over five days and would reach the moon in seven weeks. A trip to the sun, however, would take fifty-three years.

After traveling for fourteen and a fraction years we would pass the orbit of Venus and eighteen years later the orbit of Mercury. If we preferred to travel outward from the earth in the direction of Mars and the outer planets instead of toward the sun, more than twenty-seven years would elapse before we would reach the orbit of Mars. An airplane journey to Jupiter would be a matter of more than two hundred years, to Saturn four hundred and fifty years, to Uranus nearly one thousand years, and to Neptune, about one thousand five hundred years. To cross the solar system on the diameter of Neptune's orbit in an airplane, traveling day and night without stopping at the rate of 200 miles per hour would take more than three thousand years. The sun's attraction reaches far beyond Neptune's orbit, however. There are comets belonging to the solar system compelled by the sun's attraction to accompany him on his travels through space that return periodically to the immediate vicinity of the sun from regions far beyond the orbit of Neptune and there is also the possibility that one or more undiscovered planets may travel around the sun in orbits far exterior to Neptune's orbit.

Measured in terms of familiar units, such as are employed for the measurement of distances on our own planet, the extent of the solar system is tremendously great. Viewed from Neptune, the sun is so far away that it presents no appreciable disk. It is in this sense star-like to the Neptunians, but at the distance of Neptune the stars appear no more brilliant and no nearer than they do to us.

To Neptune the sun, though star-like in form, supplies a very appreciable quantity of light and heat (one nine-hundredth of the amount the earth receives) while the amount of light and heat that Neptune receives from the nearest stars is entirely inappreciable. When our airplane reaches Neptune after a journey of one thousand five hundred years, it is, as it were, just clearing the ground for its flight to the stars. To cover the intervening space to the nearest star, traveled by light in four and a third years, an airplane would need fourteen and a half million years. In that time the solar system itself would be in some far distant part of the universe, since it is speeding onward through space at the rate of twelve miles a second or about four astronomical units a year.

Changing now our unit of measurement that we may express interstellar distances in comprehensible numbers, we prepare to travel from the earth to the stars with the velocity of light.

With this velocity, one hundred and eighty-six thousand miles per second, we circumnavigate our globe in one-seventh of a second, reach the moon in one and a fourth seconds and the sun in eight minutes. In a little over four hours we pass the orbit of Neptune and are started on our journey to the stars, penetrating further and further into interstellar space. For a year we travel and reach not a single star though we are speeding ever onward with the velocity of light. We have now covered the distance of one light-year, which means that the waves of light from the sun we have left behind must travel for a year before they reach us. We continue our journey and find ourselves next at a distance of one parsec from the sun. We have traveled a distance of approximately three and a quarter light-years, and were it possible to see the earth as well as the sun at this distance, the two would appear to be but one second of arc apart, a distance that requires the most careful adjustment and manipulation of the telescope to measure accurately. We are still one light-year distant from Alpha Centauri, the nearest of the bright stars. A few of the stars will now appear somewhat brighter than they appeared to us on earth, but the majority of the stars appear just as we see them here and the forms of the constellations remain practically unchanged in appearance, for we are only beginning our journey through the sidereal universe and our position in it has only shifted by a very slight amount. If we should continue our journey to the immediate vicinity of Alpha Centauri, we would find that it is not like our own sun, a single star, but is a binary star consisting of two suns in revolution around their common center of gravity. The distance of this binary system from the solar system has been measured with considerable accuracy and is known to be four and a third light-years. Though there may be a few faint stars or non-luminous stars nearer to us than Alpha Centauri, this star has long held the distinction of being the nearest of the stars. As the sun continues his journey through the universe the two stars, Alpha Centauri and our sun, will finally draw away from each other after many ages have passed and some other sun of space will be our nearest star. The distances that separate the stars from each other probably average as great as the distance from the sun to Alpha Centauri. Within a sphere whose center is at the earth and whose radius is five parsecs, or about sixteen light-years, there are only about twenty known stars. There is, therefore, small chance of collision among bodies that are so small in proportion to the tremendous intervals of space that separate them from each other. There is ample room for the individual stars to pursue their journey through space without interfering with each other's motion so long as they are as widely scattered as they appear to be in this portion of the universe. The fact that our own sun has continued its journey through the universe for some hundreds of millions of years without any catastrophe such as would result from closely approaching or colliding with another sun of space shows how enormous is the scale upon which our sidereal system is fashioned.

Stars that are ten, fifty or even one hundred light-years from the earth are our nearest neighbors in space. They are the stars that show a slight displacement in the heavens or measurable parallax, viewed from opposite sides of the earth's orbit. There are probably a thousand stars among the hundreds of millions of stars within reach of the greatest telescopes whose distances have been determined in light-years by direct measurement of their displacement in the heavens resulting from the change of position of the earth in its orbit. The most distant of the stars are apparently immovable in the heavens showing neither the effect of the sun's motion or their own motion through space. Methods for finding the distances of many far remote stars and star-clusters have been devised, however, and some comparatively recent investigations have given results for the distances of these objects indicating that the diameter of the system of stars to which our sun belongs is approximately three hundred thousand light-years. It is difficult to grasp the full significance of this fact. It means that hundreds of millions of the suns of space throng the visible universe at distances from us and from each other running into hundreds, thousands and even hundreds of thousands of light-years. The light waves from some tiny object that we view today in one of our great reflectors may have started on their journey through space over one hundred thousand years ago when men of the Old Stone Age inhabited our planet earth!

Astronomers have found as a result of their investigations that the sidereal system to which our solar system belongs is in the form of a flattened spheroid with its longest axis in the plane of the Milky Way. The extent of this star system composed of hundreds of millions of individual suns in addition to nebulæ and clusters is probably something like three hundred thousand light-years along its longest axis, while globular star clusters lying above and below its central plane are estimated to be at distances from it ranging from ten thousand to two hundred thousand light-years. This entire organized system is our sidereal universe. Space beyond is unexplored. The globular star clusters are among the most distant celestial objects so far discovered. The spiral nebulæ may be entirely within the limits of this system or they may be even more distant than the globular clusters for their distances are not known as yet.

There is a possibility that our sidereal universe, vast as it is known to be, may be but a unit in some still greater unit and that other similar systems lie beyond the reach of existing telescopes at unimaginable distances.

The mind of man is overwhelmed by the thought of sidereal systems as vast as our own lying far beyond his ken. Whether or not such external systems do exist and are with our own sidereal system units in some still vaster creation we cannot know.

So vast, indeed, is this one visible universe of ours that the mind of man, accustomed to earthly standards, cannot comprehend its magnitude or the infinitesimal size of our whole solar system compared to it.

## CHAPTER XXXI. SOME ASTRONOMICAL FACTS WORTH REMEMBERING.

Kepler's Three Laws of Planetary Motion:

I. The planets move in ellipses with the sun at one focus.

II. The radius vector of a planet (line adjoining sun and planet) sweeps over equal areas in equal times.

III. The square of the time of revolution (the year) of each planet is proportional to the cube of its mean distance from the sun.

Sir Isaac Newton discovered that the law of gravitation extends to the stars. That is, every mass in the universe attracts every other mass with an attraction directly proportional to the product of the masses and inversely proportional to the square of the distances between them.

Ocean tides are caused by the difference between the attraction of the sun and moon for the main body of the earth and their attraction for different particles of the earth's surface. The tide-raising force of the disturbing body is proportional to its mass and inversely proportional to the cube of its distance. The tides produced by the sun are, therefore, only two-fifths as great as the tides produced by the moon.

The celestial sphere is an imaginary sphere of infinite radius, with the earth at its center, upon which the celestial bodies are considered to be projected for convenience in determining their positions with respect to fixed points of reference in the heavens.

The north and south poles of the heavens are the points on the celestial sphere directly above the north and south poles of the earth.

The celestial equator is the great circle in which the plane of the earth's equator intersects the celestial sphere. It passes through the east and west points of the horizon and through the zenith—or point directly overhead—at the earth's equator.

The ecliptic is the great circle in which the plane of the earth's orbit intersects the celestial sphere. The celestial equator and the ecliptic are inclined to each other at an angle of 23½°, which is called the obliquity of the ecliptic. The two points in which the celestial equator and the ecliptic intersect are called respectively the vernal equinox and the autumnal equinox.

The vernal equinox is an important point of reference on the celestial sphere.

As the position of a point on the earth's surface is determined by its longitude and latitude so the position of an object on the celestial sphere—star, sun, planet—is determined by its Right Ascension and Declination.

The Declination of a celestial object is its distance north or south of the celestial equator, measured in degrees, minutes and seconds of arc, on a great circle of the celestial sphere passing through the object and north and south poles of the heavens. These great circles are called hour circles and they correspond to the meridians or circles of longitude on the earth's surface. The declination of an object in the heavens corresponds to the latitude of a point on the earth's surface. The Right Ascension of a point on the celestial sphere corresponds to the longitude of a point on the earth's surface. It is measured—as longitude is measured—in degrees, minutes and seconds of arc or in hours, minutes and seconds of time—eastward along the celestial equator from the hour circle passing through the vernal equinox to the foot of the hour circle passing through the object. The hour circle passing through the vernal equinox is the zero meridian for the celestial sphere just as the meridian of Greenwich is the zero meridian on the earth's surface.

The mean distance of the earth from the sun is 92,900,000 miles and is called the astronomical unit.

The sun with its satellites advances through the universe at the rate of 4 astronomical units in a year or approximately one million miles a day.

The parallax of a star is the angle at the star subtended by the radius of the earth's orbit, 92,900,000 miles, or the astronomical unit. It is, in other words, the angular distance between the earth and sun as viewed from the star. The larger the parallax the nearer the star. The largest known stellar parallax is that of Alpha Centauri and its value is 0".75.

The light-year is the distance that light travels in one year. It is equal to about 63,000 astronomical units or nearly six trillion (6,000,000,000,000) miles. The velocity of light is 186,000 miles per second.

The parsec is equal to 3.26 light-years. It is the distance of a star that has a parallax of one second of arc.

The apparent magnitude of a star is its apparent brightness estimated on a scale in which a difference of one magnitude corresponds to a difference in brightness of 2.51, or the fifth root of one hundred. A difference of five magnitudes corresponds to a difference one one hundredfold in brightness, of ten magnitudes to ten thousandfold in brightness. In exact measurements on this scale magnitudes are estimated to tenths.

Stars that are one magnitude brighter than stars of the standard first magnitude are of the zero magnitude and stars still brighter are of negative magnitudes.

Sirius is a star of the -1.6 magnitude. Jupiter at opposition is of -2.0 magnitude and Venus at greatest brilliancy of -4.0 magnitude. The sun on this scale of comparative brightness is of the -26.7 apparent magnitude. The faintest stars visible in the most powerful telescope in the world—the 101-inch Mt. Wilson Hooker telescope—are of the twentieth magnitude.

The absolute magnitude of a star is its apparent magnitude at the standard distance of ten parsecs or 32.6 light years. The absolute magnitude of the sun is five. That is, the sun would be a fifth-magnitude star at the standard distance of 32.6 light-years. The absolute magnitudes of stars indicate how bright they would be relatively if they were all at the same standard distance. Apparent magnitudes indicate how bright the stars appear to be at their true distances.

The mean distance of the moon from the earth is approximately 240,000 miles or sixty times the earth's radius.

The sun is four hundred times farther away than the moon and its diameter is about four hundred times greater than the moon's diameter.

The nearest star is about 275,000 times more distant than the sun, and the most distant known object, the globular star cluster, N.G.C. 7106, is about fourteen billion times more distant than the sun.

The earth is a spheroid flattened at the poles and its polar diameter is about twenty-seven miles shorter than its equatorial diameter. An object weighs less at the poles than at the equator.

The earth's interior is as rigid as steel and probably consists of a core of magnetic iron surrounded by an outer stony shell.

Eclipses of the sun occur when the moon passes between the earth and sun. They can only occur at the time of new moon. There must be at least two solar eclipses every year separated by an interval of six months and there may be as many as five solar eclipses in a year. Eclipses of the moon occur when the earth comes between the sun and moon, and the moon passes into the earth's shadow. Eclipses of the moon can only occur at full moon. There may or may not be eclipses of the moon every year. The greatest number of eclipses than can occur in any one year, solar and lunar combined, is seven and the least number is two and in that case they are both solar eclipses.

The sun is a yellow, dwarf star of a density of one and one-fourth that of water and with a surface temperature of about 12,000° F. except in sun-spot regions where the temperature is about 6,000° F. It is probably gaseous throughout.

The sun, as well as the planets, rotates on its axis and different portions of the surface rotate at slightly different rates. The average period of the rotation of the sun on its axis is about twenty-six days.

The sun is a variable star with a twofold variation. One is of long period during the eleven-year sun-spot cycle with a range of from three to five per cent. The other is a short irregular variation with a period of a few days, weeks or months and a range of from three to ten per cent.

Sun-spots are solar cyclones and appear black only by contrast with their hotter and brighter surroundings. They come in eleven-year cycles (approximately) with periods of maximum and minimum appearance.

The brightness and blue color of the sky is due to the scattering of sunlight by the molecules of oxygen and nitrogen in the earth's upper atmosphere. If there were no atmosphere the skies would appear black except in the direction of the heavenly bodies, which would be visible by day as well as by night.

The solar corona is the rare outer envelope of the sun and it is visible only during a total eclipse of the sun. It is partly of an electrical nature and it varies in form during the sun-spot cycle. It often extends to a distance of several solar diameters on either side of the sun.

The warmth and the habitability of the earth's surface is due to the presence of water-vapor and carbon-dioxide in the atmosphere. Without these substances in the atmosphere life on the earth's surface would be impossible.

Half of the earth's atmosphere and all clouds lie within seven miles of the earth's surface, and at high elevations above the earth the temperature is many degrees below zero.

The temperature of space approaches the absolute zero of -459° F.

The only planets in the solar system with the exception of the earth that might possibly support life are Venus and Mars.

Stars shine by their own light but planets shine only by reflected light from the sun.

If the earth were represented by a six-inch school globe the sun would be on the same scale a globe fifty-four feet in diameter. Mercury would be a small ball two and a third inches in diameter. Venus would be another six-inch globe. Mars would be a ball about the size of a baseball, three and a fifth inches in diameter. The moon would be about the size of a golf ball, one and a half inches in diameter. The largest asteroids would be the size of marbles. Average-sized asteroids would be the size of shot and the smallest would be merely grains of sand.

Jupiter would be a huge globe standing as tall as a man five feet six inches in height. Saturn would be a smaller globe four and a half feet in diameter and its ring system would extend to a distance of five and a half feet on either side of the globe. Uranus would be represented by a globe almost exactly two feet in diameter and Neptune would be a slightly larger globe with a diameter of two feet two and a half inches.

The satellites of the outer planets would range in size from tennis and golf balls for the largest, to marbles for the smaller and grains of sand for the smallest.

On the same scale of measurement the distance of the six-inch globe of the earth from the fifty-four foot globe representing the sun would be one and one-tenth miles. The moon would be placed fifteen feet from the earth-globe and the diameter of the solar system on the same scale measured across the orbit of Neptune would be sixty-six miles. The nearest star on this scale would be three hundred thousand miles away.

If the distance from the earth to the sun is taken as one inch so that the scale of the universe is reduced six trillion times, the diameter of the solar system across Neptune's orbit is five feet and the distance of one light-year comes out almost exactly equal to one mile. The nearest star to the five-foot solar system would be four and a third miles away; the most distant known object would be two hundred and twenty thousand miles away, and the extent of the visible universe would be three hundred thousand miles. On the same scale the diameter of our sun would be about one hundredth of an inch and the diameters of the giant stars Antares and Betelgeuze would be four inches and two and three-fourth inches respectively. To see the earth we would need a microscope.